Adaptive Control of Dynamic Systems with Uncertainty and Quantization

Automation and Control Engineering
Series Editors - Frank L. Lewis, Shuzhi Sam Ge, and Stjepan Bogdan

Optimal and Robust Scheduling for Networked Control Systems
Stefano Longo, Tingli Su, Guido Herrmann, and Phil Barber

Electric and Plug-in Hybrid Vehicle Networks
Optimization and Control
Emanuele Crisostomi, Robert Shorten, Sonja Stüdli, and Fabian Wirth

Adaptive and Fault-Tolerant Control of Underactuated Nonlinear Systems
Jiangshuai Huang, Yong-Duan Song

Discrete-Time Recurrent Neural Control
Analysis and Application
Edgar N. Sánchez

Control of Nonlinear Systems via PI, PD and PID
Stability and Performance
Yong-Duan Song

Multi-Agent Systems
Platoon Control and Non-Fragile Quantized Consensus
Xiang-Gui Guo, Jian-Liang Wang, Fang Liao, Rodney Swee Huat Teo

Classical Feedback Control with Nonlinear Multi-Loop Systems
With MATLAB® and Simulink®, Third Edition
Boris J. Lurie, Paul Enright

Motion Control of Functionally Related Systems
Tarik Uzunović and Asif Sabanović

Intelligent Fault Diagnosis and Accommodation Control
Sunan Huang, Kok Kiong Tan, Poi Voon Er, and Tong Heng Lee

Nonlinear Pinning Control of Complex Dynamical Networks
Edgar N. Sanchez, Carlos J. Vega, Oscar J. Suarez, and Guanrong Chen

Adaptive Control of Dynamic Systems with Uncertainty and Quantization
Jing Zhou, Lantao Xing, and Changyun Wen

For more information about this series, please visit: https://www.crcpress.com/ Automation-and-Control-Engineering/book-series/CRCAUTCONENG

Adaptive Control of Dynamic Systems with Uncertainty and Quantization

Jing Zhou

Lantao Xing

Changyun Wen

CRC Press
Taylor & Francis Group
Boca Raton London New York

CRC Press is an imprint of the
Taylor & Francis Group, an **informa** business

First edition published 2022
by CRC Press
6000 Broken Sound Parkway NW, Suite 300, Boca Raton, FL 33487-2742

and by CRC Press
2 Park Square, Milton Park, Abingdon, Oxon, OX14 4RN

© 2022 Jing Zhou, Lantao Xing, Changyun Wen

CRC Press is an imprint of Taylor & Francis Group, LLC

ISBN: 978-1-032-00981-0 (hbk)
ISBN: 978-1-032-00982-7 (pbk)
ISBN: 978-1-003-17662-6 (ebk)

DOI: 10.1201/9781003176626

Publisher's note: This book has been prepared from camera-ready copy provided by the authors

Typeset in Nimbus Roman
by KnowledgeWorks Global Ltd.

Contents

SECTION III: INPUT AND STATE/OUTPUT QUANTIZATION COMPENSATION   147

SECTION IV: APPLICATIONS   177

Preface

In modern control engineering systems, sensing and communication technologies are commonly used and have broad practical applications, such as digital control systems, hybrid systems, electrical power grid systems, intelligent transportation systems, and robotic networks. These control systems are implemented via communication channels with limited information, such as quantized signals. Due to the theoretical and practical importance of the study of such dynamic systems, there has been a great deal of interest in the development of advanced control of quantized dynamic systems.

In such control engineering applications, quantization is not only inevitable owing to the widespread use of digital processors that employ finite-precision arithmetic, but also useful. An important aspect is to use quantization schemes that yield sufficient precision, but require a low communication rate. However, quantization will introduce strong nonlinear and discontinuous characteristics to the system and will destroy the performance, and may even make the closed-loop system unstable. A direct application of the existing tools to quantized control design will not yield satisfactory results, particularly in nonlinear systems. If the system has uncertainty or the system parameters are poorly known, the compensation problem will become more complicated. Though adaptive control has been proved to be a promising tool to control uncertain nonlinear systems, several important issues, such as compensation of effects of quantized input signals and/or quantized output signals and guaranteeing the transient performance of adaptive compensation control systems, remain less explored.

In this book, a series of innovative technologies and research results on adaptive control of dynamic systems with quantization, uncertainty, and nonlinearity are presented, including the theoretical success and practical development such as the approaches for stability analysis, the compensation of quantization, the treatment of subsystem interactions, and the improvement of system tracking and transient performance. Compared with the existing literature, novel solutions by adopting backstepping design tools to a number of hotspots and challenging problems in the area of adaptive control are provided. The results are given in three parts.

In the first part, adaptive backstepping based control schemes will be introduced to compensate for the effects of input quantization and to solve the problems of stabilization of nonlinear uncertain systems with input quantization (Chapter 4), tracking control of uncertain nonlinear systems with input quantization (Chapter 5), decentralized control of interconnected systems with inputs quantization (Chapter 6), and output feedback control of nonlinear uncertain systems with input quantization (Chapter 7). It will be shown how the effects of input quantization can be compensated and how the desired system performance is achieved, by incorporating the backstepping technique with other methodologies, such as introducing hyperbolic function in the controller, decentralized control, and observer design. The proposed adaptive control schemes are shown to ensure the stability of the resulting control system. With these schemes, system performances can be precisely characterized as functions of design parameters and thus are tunable in a certain sense by designers.

The second part involves designing and analyzing adaptive backstepping controllers for a class of uncertain nonlinear systems with state quantization, including bounded quantizers (Chapter 8) and sector-bounded quantizers (Chapter 9). Newly developed stability analysis strategies are presented by constructing a new compensation scheme for the effects of the state quantization and handling discontinuity resulting from the state quantization. The developed controllers are shown to guarantee the ultimate boundedness of the closed-loop system and the error performance is established and can be improved by appropriately adjusting design parameters.

In the third part, two different adaptive compensation methods will be introduced for solving the control problem of simultaneous input and state/output quantization for uncertain nonlinear systems. Issues including adaptive backstepping state feedback control of both input and state quantization (Chapter 10) and adaptive output feedback control of both input and output quantization (Chapter 11) are discussed in detail.

In the last part, the developed adaptive control and compensation schemes are applied to helicopter and DC microgrid with quantized signals.

Discussion remarks are provided in each chapter highlighting new approaches and contributions to emphasize the novelty of the presented design and analysis methods. In addition, simulation results are given in each chapter to show the effectiveness of these methods.

This book is helpful in learning and understanding the fundamental backstepping schemes for state feedback control and output feedback control. It can be used as a reference book or a textbook on adaptive quantized control for students with some background in feedback control systems. The book is also intended to introduce researchers and practitioners to the area of adaptive control systems involving the treatment of input and/or state quantization, nonlinear functions, uncertain parameters, and interactions. Researchers, graduate students, and engineers in the fields of control, information, and communication, electrical engineering, mechanical engineering, applied mathematics, computer science, and others will benefit from this book.

We would like to express our deep sense of gratitude to our beloved families who have made us capable enough to write this book. Jing Zhou is greatly indebted to her husband Xiaozhong Shen, her children Zhile Shen, Arvid Zhiyue Shen and Lily Yuxin Shen, and her parents Feng Zhou and Lingfang Ma, for their care, under-standing and constant support throughout these years.

Lantao Xing is greatly indebted to his wife Yue Li, his son Shumu Xing, and his beloved parents for their constant support throughout these years.

Changyun Wen is greatly indebted to his wife Xiu Zhou and his children Wen Wen, Wendy Wen, Qingyun Wen and Qinghao Wen for their constant invaluable support and assistance throughout these years.

The authors would like to thank the series editor, Sam Ge and Frank Lewis, and the entire team of CRC Press for their cooperation and great efforts in bringing to fruition the work in the book.

Finally, the authors are grateful to the University of Agder (Norway), Nanyang Technological University (Singapore) and Zhejiang University (China), for providing plenty of resources for our research work. The research presented in this book was supported partly by the Research Council of Norway under Grant 306640 (DEEP-COBOT Project) and Grant 309582 (INMOST Project).

Authors

Jing Zhou received her BEng from Northwestern Polytechnical University, China in 2000, and PhD from Nanyang Technological University, Singapore in 2006. She was a senior research scientist at International Research Institute of Stavanger (NORCE) in Norway from 2009–2016 and a postdoctoral fellow at Norwegian University of Science and Technology from 2007–2009, respectively. Since 2016, she has been with the Faculty of Engineering and Science, University of Agder, Norway, where she is currently a full professor and a research director of the Priority Research Center of Mechatronics. Her research interests are in the fields of adaptive and nonlinear control, networked control systems, cyber-physical systems, robotics and vision, and control applications to offshore mechatronics systems including cranes, marine vessels, industry robots, aircraft, and drilling and well systems. She has published 2 books and over 100 papers in international refereed journals and conferences. Prof. Zhou is a fellow of Norwegian Academy of Technological Sciences (NTVA). She is currently the associate editor of *IEEE Transactions on Cybernetics, Systems & Control Letters*, and IEEE CSS Conference. She also serves as an IEEE CSS technical committee on "Nonlinear Systems and Control" and "System Identification and Adaptive Control". She has been actively involved in organizing international conferences playing the roles of General Chair, General Co-Chair, Technical Program Committee Chair, Program Committee Member, Invited Session Chair, etc.

Lantao Xing received his BEng in automation from China University of Petroleum (East China), China, in 2013, and PhD in Control Science and Engineering from Zhejiang University, China, in 2018. Then he worked as a research fellow with the School of Computer Science, Queensland University of Technology, Australia. Currently, he is a Presidential Postdoctoral Fellow at the School of Electrical and Electronic Engineering, Nanyang Technological University, Singapore. His research interests include nonlinear system control, event-triggered and quantized control, and distributed control with applications in smart grid.

Changyun Wen earned a BEng from Xi'an Jiaotong University, China in 1983 and a PhD from the University of Newcastle, Australia in 1990. From 1989–1991, he was a postdoctoral fellow at University of Adelaide, Australia. Since 1991, he has been with Nanyang Technological University, Singapore where he is currently a full professor. His main research activities are in the areas of control systems and applications, cyber-physical systems, smart grids, complex systems and networks. Prof. Wen is a fellow of IEEE, was a member of the IEEE Fellow Committee from 2011–2013 and a Distinguished Lecturer of IEEE Control Systems Society from 2010–2013. He is currently the co-editor-in-chief of *IEEE Transactions on Industrial Electronics*, associate editor of *Automatica* (from 2006) and executive editor-in-chief of *Journal of Control and Decision*. He also served as an associate editor of *IEEE Transactions on Automatic Control* from 2000–2002, *IEEE Transactions on Industrial Electronics* from 2013–2020 and *IEEE Control Systems Magazine* from 2009–2019. He has been actively involved in organizing international conferences playing the roles of General Chair (including the General Chair of IECON 2020 and IECON 2023) and TPC Chair (e.g., the TPC Chair of Chinese Control and Decision Conference since 2008). He was the recipient of a number of awards including the Prestigious Engineering Achievement Award from the Institution of Engineers, Singapore in 2005, and the Best Paper Award of *IEEE Transactions on Industrial Electronics* in 2017.

Chapter 1

Introduction

Due to the rapid development of sensing and communication technologies, there have been great efforts to develop high-performance control schemes for dynamic plants like electric power grid systems, intelligent transportation systems, and robotic networks. There has been a great deal of interest to deal with the fundamental system characteristics, such as uncertainty and nonlinearity, and the information limits, such as quantized signals. Adaptive control has been proved to be one of the most promising techniques which can be applied to control a wide variety of systems and processes. Adaptive control theory attempts to improve the behavior or performance of physical systems by gathering and exploiting knowledge about the system's operation. Usually, this knowledge is encoded as a descriptive mathematical model of the physical plant from which the controller design is derived. Given a mathematical representation, there are great interests in designing adaptive controllers for quantized control systems to achieve objectives such as stability (convergence of state), output tracking of some reference signals, and transient performance. Although adaptive control has been proved to be a promising tool to control uncertain systems, several important issues, such as adaptive compensation of input/state/output quantization, stability analysis of adaptive compensated systems, and guaranteeing the transient performance of corresponding closed-loop systems, still have not been extensively explored.

1.1 Adaptive Control

Adaptive control has been an important area of active research for over six decades. Significant development has been seen, including theoretical success and practical development, such as the proof of global stability and the improvement of

DOI: 10.1201/9781003176626-1

system tracking and transient performance. One of the reasons for extensive research activities and the rapid growth of adaptive control is its ability to control plants with uncertainties during its operation. Adaptive control is a technique of applying some methods to obtain a real-time model of the process online and using this model to design a controller. An adaptive controller is designed by combining a parameter estimator, which provides estimates of unknown parameters, with a control law. The parameters of the controller are adjusted during the operation of the plant. To obtain desired performances, it also provides adaptation methods to deal with some uncertainties, such as flow and speed variations, external disturbance, and structural uncertainties. One important approach in adaptive control is certainty equivalence-based design. Such an approach has been studied extensively and a number of results have been established [39, 84–88, 97]. Certain schemes have also been proposed to study the robustness issues in the context of both single loop control [40, 42, 72–74, 106, 107, 111] and decentralized control of multi-loop systems [18, 19, 35, 38, 41, 110, 112–114].

At the beginning of the 1990s, a new approach called "backstepping" was proposed for the design of adaptive controllers. The technique was comprehensively addressed by Krstic, Kanellakopoulos and Kokotovic in [53]. Backstepping is a recursive Lyapunov-based scheme for the class of "strict feedback" systems. In fact, when the controlled plant belongs to the class of systems transformable into the parametric-strict feedback form, this approach guarantees global or regional regulation and tracking properties. An important advantage of the backstepping design method is that it provides a systematic procedure to design stabilizing controllers, following a step-by-step algorithm. With this method, the construction of feedback control laws and Lyapunov functions is systematic. Another advantage of backstepping is that it has the flexibility to avoid cancellations of useful nonlinearities and achieve stabilization and tracking. A number of results using this approach have been obtained [21, 37, 64, 70, 104, 108, 115–119, 134–141, 143, 144, 148–150, 152, 153]. However, there are still some important unaddressed issues such as compensating for the effects of input quantization, state quantization, and output quantization in unknown dynamic systems.

1.2   Motivation

In modern control engineering systems, information, sensing, and communication technologies are commonly used and have broad practical applications, such as electric power grid systems, intelligent transportation systems, computer networks, and robotic networks. When dealing with control engineering problems, the designer is inevitably led to face the difficulties tied to limited information, such as quantization. Quantization technique is widely used in digital control, hybrid systems, networked systems, signal processing, simulation, embedded computing, and so on. It is useful and inevitable to minimize information flow, decrease communication burden, and increase system security.

Quantization introduces strong nonlinear characteristics to the system, such as discontinuity, and this may degrade the control performance or even make the closed-loop system unstable. A direct application of the existing tools to quantized control design will not yield satisfactory results, particularly for nonlinear uncertain systems. There are a great amount of theoretical and practical interests in addressing this problem and developing new theories that capture both controller design and quantization effects. As a result, quantized control has attracted considerable attention in recent years. Certain design methods based on different control objectives and system conditions have been developed and verified in both theory and practice.

1.2.1 Control of Input Quantization

Research on stabilization of linear and nonlinear systems with quantized control signals has received great attention, see for examples, [23, 45, 61, 62, 75, 81, 98]. The systems considered in the above references are completely known. In practice, it is often required to consider the case where the plant to be controlled is uncertain. Quantized control of systems with uncertainties has been studied by using robust approaches, see for examples, [20, 66, 67, 67, 80, 129]. As well known, adaptive control is a useful and important approach to deal with system uncertainties due to its ability to provide online estimations of unknown system parameters with measurements. It is noted that adaptive control schemes with quantized input have been reported in [33, 34, 95]. In [33, 95], adaptive control for linear uncertain systems with input quantization was studied. In [34], adaptive quantized control of nonlinear systems was considered, where the idea of constructing the hysteretic type of input quantization was originally introduced. However, the stability condition in [33, 34] depends on the control signal, which is hard to be checked in advance as the control signal is only available after the controller is put in operation.

Due to a number of advantages of the backstepping technique, such as providing a promising way to improve the transient performance of adaptive systems by tuning design parameters, adaptive backstepping control of uncertain nonlinear systems with quantized input has been studied in [124, 125, 127, 142, 145, 147]. In [142, 147], the stabilization of uncertain nonlinear systems with input quantization is considered. Although the proposed method can relax the stability condition in [33, 34], it requires the nonlinear functions to satisfy global Lipschitz conditions with known Lipschitz constants. Also, the proposed controller must follow a guideline to select the parameters of the quantizer to ensure the stability of the closed-loop system. Such Lipschitz conditions are relaxed in [124] and [145]. In [124], an implicit adaptive controller was developed for the system where unknown parameters only appear in the last differential equation of the system and the controller is contained in an equation, that is, related to a hyperbolic tangent function. It is not easy to solve such an equation to get an explicit controller. In [145], a hyperbolic function is introduced into the adaptive controller to compensate for the effects of input quantization. Similarly, a novel smooth function is adopted in [59] to generate the controller which can eliminate the effects of input quantization and actuator faults. In [125] and [127], the problem of output feedback control for uncertain nonlinear systems with input quantization was addressed, where state observers were designed to estimate the unknown states.

In the control of uncertain interconnected systems, decentralized adaptive control strategy, designed independently for local subsystems and using only locally available signals for feedback propose, is an efficient and practical strategy. In the control of interconnected systems with input quantization, the number of available decentralized control is still limited. In [101], the issue of decentralized quantized control via output-feedback for interconnected systems has been addressed, where the original system needs to be transformed to a form including only the output signal and the signals from filters. In this way, interactions only exist in the equation related to the output signal in the final controlled systems and the rest equations related to the filter signals do not involve interactions. In [145], a more general class of interconnected systems with input quantization is considered in the sense that interactions exist in all the differential equations of the subsystems. In the above paper, a totally decentralized adaptive controller design approach is developed together with a new compensation method constructed for the unknown nonlinear interactions and quantization error.

1.2.2   Control of State Quantization

For a control system with state quantization, the state measurements are processed by quantizers, which are discontinuous maps from continuous spaces to finite sets. Such discontinuous property may lead to the control design and stability analysis difficulty. Feedback control of systems with state quantization has attracted growing interest lately in [60, 63, 67], where the systems considered are completely known. Uncertainties and nonlinearities always exist in many practical systems. Thus, it is more reasonable to consider controller design for uncertain nonlinear systems. Although adaptive control of uncertain systems has received considerable interest and been widely investigated, there are still limited works devoted to adaptive control with state quantization. It is noted that adaptive control schemes for linear systems with state quantization have been reported only in [3, 99]. In [99], a supervisory control scheme for uncertain linear systems with quantized measurements has been proposed. While in [3], the adaptive control of linear systems with quantized measurements and bounded disturbances has been addressed.

Research on adaptive control of uncertain systems with state quantization using the backstepping technique is still limited. The involved major difficulty is that the backstepping technique requires differentiating virtual controls and in turn the states by applying the chain rule. If the states are quantized, they become discontinuous, and therefore, it is difficult to analyze the resulting control system with the current backstepping-based approaches. Recently, an effective adaptive backstepping control scheme was proposed for uncertain nonlinear systems with state quantization in [146], where the quantization error of the states must be bounded by a constant. By using backstepping approaches, a new adaptive control algorithm using only quantized states is developed by constructing a new compensation method for the effects of state quantization.

1.2.3   *Control of Both Input and State/Output Quantization*

It is common in networked control systems that the sensor and control signals are transmitted via a common communication network. In such a quantized control system, the state measurements and the control signals are both processed by quantizers. So far, only a few works have been reported to handle the issue with the simultaneous existence of quantizers in both uplink and downlink communication channels of control systems. References [7, 8, 17, 82, 131] are some examples, however, only linear systems are considered. [126] studied adaptive output feedback regulation for uncertain nonlinear systems with input and output quantization. Research on adaptive control of uncertain nonlinear systems with both state and input quantization using backstepping technique is still limited. The main challenge is that only quantized states can be utilized to construct the virtual controls in each recursive step. Hence, the virtual controls are discontinuous, of which the derivatives cannot be computed as often done in standard backstepping design procedure.

1.3   Objectives

In this book, a series of innovative technologies and research results on adaptive control of dynamic systems with quantization and uncertainty are presented, including the theoretical success and practical development such as the approaches for stability analysis, the compensation of effects of quantization, the treatment of subsystem interactions, and the improvement of system tracking and transient performance. Compared with the existing literature, novel solutions by adopting backstepping design tools to a number of hotspots and challenging problems in the area of adaptive control are provided.

The main objectives of this book are listed as follows:

■ The first part of the book is to introduce the backstepping design to compensate for the effects of input quantization in nonlinear uncertain systems in both state feedback control (Chapters 4–6) and output feedback control (Chapter 7). Issues include stabilization of nonlinear uncertain systems with input quantization (Chapter 4), tracking control of nonlinear uncertain systems with input quantization (Chapter 5), decentralized control of interconnected systems with inputs quantization (Chapter 6), and output feedback control of nonlinear uncertain systems with input quantization (Chapter 7). It will be shown how the effects of input quantization can be compensated and how the desired system performance is achieved, by incorporating the backstepping technique with the other methodologies, such as introducing hyperbolic function in the controller, decentralized control, and observer design. The proposed adaptive control schemes are shown to ensure the stability of the resulting control system. With these schemes, system performances can be precisely characterized as functions of design parameters and thus is tunable in certain sense by designers.

■ The second part of the book is to control a class of uncertain nonlinear systems with state quantization, including bounded quantizers (Chapter 8) and sector-bounded quantizers (Chapter 9). The aim is at designing, analyzing, and implementing adaptive backstepping control methods which can accommodate quantized states by introducing new adaptive compensation schemes to overcome or compensate for the effect of state quantization and uncertainty. Newly developed stability analysis strategies are presented to compensate for the effects of the state quantization and handle discontinuity resulted from the state quantization. The developed controllers are shown to guarantee the ultimate boundedness of the closed-loop system. Moreover, the error performance is established and can be improved by appropriately adjusting design parameters.

■ In the third part, two different adaptive compensation methods using backstepping design will be introduced for solving the control problem of simultaneous input and state/output quantization for uncertain nonlinear systems. Issues including adaptive state feedback control of both input and state quantization (Chapter 10) and adaptive output feedback control of both input and output quantization (Chapter 11) are discussed in detail.

■ In the last part of the book, the developed control strategies are successfully applied to two practical systems: helicopter system with quantized states (Chapter 12) and electrical power system with quantized signal transmission (Chapter 13).

1.4   Preview of Chapters

This book is divided into fourteen chapters. Chapters 2–14 are previewed below.

In Chapter 2, the concepts of adaptive backstepping control and robust backstepping control designs, as the basic tool of new contributions achieved in the remaining chapters, are given. Recursive design procedures using both state-feedback and output feedback are presented. Approaches of establishing system stability and performances are also given.

In Chapter 3, four types of quantized control systems are presented including a system with input quantization, a system with state quantization, a system with input and state quantization, and a system with input and output quantization. The basic descriptions and properties of five quantizers are presented, including uniform quantizer, logarithmic quantizer, hysteresis-logarithmic quantizer, hysteresis-uniform quantizer, and logarithmic-uniform quantizer.

In Chapter 4, adaptive stabilization for a class of uncertain nonlinear systems in the presence of input quantization is investigated. The considered nonlinear systems satisfy the Lipschitz continuity condition and a hysteresis-logarithmic quantizer is considered. The control design is achieved by using the backstepping technique and a guideline is derived to select the parameters of the quantizer. The designed controller together with the quantizer ensures the stability of the closed-loop system in the sense of signal boundedness.

In Chapter 5, adaptive tracking control for a class of uncertain nonlinear systems in the presence of input quantization is developed. Several types of quantizers are considered in this chapter, including the uniform quantizer, the logarithmic quantizer, and the hysteresis-logarithmic quantizer. By using the backstepping technique, a new adaptive control algorithm is developed by constructing a new compensation method for the effects of the input quantization. A hyperbolic tangent function is introduced in the controller with a new transformation of the control signal. The developed controllers do not require the Lipschitz condition for the nonlinear functions and also the quantization parameters can be unknown. Besides showing global stability, tracking error performance is also established and can be adjusted by tuning certain design parameters.

In Chapter 6, adaptive backstepping control is employed for a class of interconnected systems with unknown interactions and with the input of each loop preceded by quantization. A totally decentralized adaptive control scheme is developed with a new compensation method incorporated for the unknown nonlinear interactions and quantization error. Each local controller, designed simply based on the model of each subsystem by using the adaptive backstepping technique, only employs local information to generate control signals. Besides showing stability, tracking error performance is also established and can be adjusted by tuning certain design parameters.

In Chapter 7, adaptive output-feedback tracking control schemes for a class of uncertain nonlinear systems with input signal quantized are presented. Two different quantizers are considered, i.e. the quantizers with bounded quantization errors and quantizers with unbounded quantization errors. The proposed schemes provide a way to relax certain restrictive conditions, in addition to solving the problem of adaptive output-feedback control with input quantization. It is shown that the designed adaptive controller ensures global boundedness of all the signals in the closed-loop system and enables the tracking error to exponentially converge towards a compact set that is adjustable.

In Chapter 8, the stabilization problem for uncertain nonlinear systems with quantized states is studied. All states in the system are quantized by a bounded quantizer, including a uniform quantizer, a hysteresis-uniform quantizer, and a logarithmic-uniform quantizer. Adaptive backstepping-based control algorithms and a new approach to stability analysis are developed by constructing a new compensation scheme for the effects of the state quantization and handle discontinuity resulted from the state quantization. Besides showing the ultimate boundedness of the system, the error performance is also established and can be improved by appropriately adjusting design parameters.

In Chapter 9, a solution for designing adaptive backstepping controllers is provided for achieving tracking of nonlinear uncertain systems in the presence of sector-bounded state quantization. By establishing the relation between the input signal and error state, the closed-loop system stability can be achieved by choosing proper design parameters.

In Chapter 10, a new adaptive backstepping control algorithm is proposed for uncertain nonlinear systems with both input and state quantization. Detailed design and analysis are given, including control structure, stability, and convergence of the algorithms. In addition to overcoming the difficulty to proceed recursive design of virtual controls with quantized states, the relation between the input signal and error states are well established to handle the effects of quantization. Thus the closed-loop system stability can be achieved without the need for sufficient conditions dependent on the design parameters and quantization parameters. It is shown that all closed-loop signals are ensured uniformly bounded and all states will converge to a compact set.

In Chapter 11, adaptive output feedback regulation for uncertain nonlinear systems, where both the output signal and the input signal of the system are quantized for the sake of less communication burden. A control law with an adaptive gain will be presented to compensate for the quantization errors. It is proved that the proposed scheme ensures that all the closed-loop signals are globally bounded. In addition, the output signal can be regulated to a bounded compact set which is explicitly given.

In Chapter 12, the developed adaptive backstepping control strategy is successfully applied to solve an attitude control problem of a helicopter system with two degrees of freedom. Only quantized input signals are used in the system which reduces communication rate and cost. Experiments are carried out on the Quanser helicopter system to validate the effectiveness, robustness, and control capability of the proposed scheme.

In Chapter 13, a quantized distributed secondary control strategy is proposed for the DC microgrid. With the help of quantizers, each converter only needs to send the quantized signals to its neighbors. It is proved that the proposed strategy can ensure proper current sharing and DC bus voltage regulation at the same time. Moreover, it also enables the DC microgrid to connect various kinds of loads.

Finally, the entire book is concluded in Chapter 14 by summarizing the main approaches and contributions and discussing some promising open problems in the areas of adaptive control in multi-agent systems, cyber-physical systems, and robotic systems.

Chapter 2

Backstepping Control

The recursive design in this book is referred to as a "backstepping design". This design method follows a step-by-step procedure because it starts with a scalar differential equation and then steps back toward the control input. The backstepping technique is a powerful tool for designing stabilizing controls for a special class of nonlinear dynamical systems. It is a recursive Lyapunov-based scheme proposed at the beginning of the 1990s. The technique was comprehensively addressed by Krstic, Kanellakopoulos, and Kokotovic in [53]. The idea of backstepping is to design a controller recursively by considering some of the state variables as "virtual controls" and designing for them intermediate control laws. Backstepping achieves the goals of stabilization and tracking. The proof of these properties is a direct consequence of the recursive procedure because a Lyapunov function is constructed for the entire system including the parameter estimates.

In the first part of the chapter, the procedures to design adaptive backstepping controllers by incorporating the tuning functions are presented for a class of parametric strict-feedback nonlinear systems with uncertainties in both state feedback and output feedback. In the second part, the procedures to design robust backstepping controllers with state feedback and output feedback are then presented for a class of strict-feedback nonlinear systems with uncertainties. The stability analysis for the adaptive backstepping and robust backstepping control schemes is also provided briefly.

2.1 Adaptive State Feedback Control

In this section, the procedures to design adaptive state-feedback backstepping controllers by incorporating the tuning functions are presented. We consider a class of

DOI: 10.1201/9781003176626-2

uncertain nonlinear systems in the following parametric strict-feedback form,

$$
\begin{aligned}
\dot{x}_1 &= x_2 + \phi_1^T(x_1)\theta + \psi_1(x_1) \\
\dot{x}_2 &= x_3 + \phi_2^T(x_1, x_2)\theta + \psi_2(x_1, x_2) \\
&\vdots \qquad \vdots \\
\dot{x}_{n-1} &= x_n + \phi_{n-1}^T(x_1, \ldots, x_{n-1})\theta + \psi_n(x_1, \ldots, x_{n-1}) \\
\dot{x}_n &= bu + \phi_n^T(x)\theta + \psi_n(x),
\end{aligned}
\tag{2.1}
$$

where $x = [x_1, \ldots, x_n]^T \in R^n$ is the state of the system, the vector $\theta \in \Re^r$ is constant and unknown, $\phi_i \in \Re^r, \psi_i \in R, i = 1, \ldots, n$ are known nonlinear functions, and the high frequency gain b is an unknown constant. It is assumed that the full states $(x_1, \ldots, x_n)$ are measurable.

The control objective is to force the output x_1 to asymptotically track the reference signal x_r with the following assumptions.

Assumption 2.1 *The sign of b is known.*

Assumption 2.2 *The reference signal x_r and its n-th order derivatives are piecewise continuous and bounded.*

2.1.1 Design of Adaptive Controllers

For system (2.1), the number of design steps required is equal to n. At each step, an error variable z_i, a stabilizing function α_i and a tuning function τ_i are generated. Finally, the control u and a parameter estimate $\hat{\theta}$ are developed.

Introduce the change of coordinates

$$
z_1 = x_1 - x_r \tag{2.2}
$$

$$
z_i = x_i - \alpha_{i-1} - x_r^{(i-1)}, \ i = 2, 3, \ldots, n, \tag{2.3}
$$

where α_i are virtual controllers. The design procedure is elaborated in the following steps.

Step 1. We start with the first equation of (2.1). The derivative of tracking error z_1 is given as

$$
\begin{aligned}
\dot{z}_1 &= \dot{x}_1 - \dot{x}_r \\
&= z_2 + \alpha_1 + \phi_1^T\theta + \psi_1.
\end{aligned}
\tag{2.4}
$$

Designing the first stabilizing function α_1 as

$$
\alpha_1 = -c_1 z_1 - \phi_1^T\hat{\theta} - \psi_1. \tag{2.5}
$$

where c_1 is a positive constant and $\hat{\theta}$ is an estimate of θ. Our task in this step is to achieve the tracking task $x_1 \to x_r$ by considering the Lyapunov function

$$
V_1 = \frac{1}{2}z_1^2 + \frac{1}{2}\tilde{\theta}^T\Gamma^{-1}\tilde{\theta}, \tag{2.6}
$$

where Γ is a positive definite matrix and $\tilde{\theta} = \theta - \hat{\theta}$. Then the derivative of V_1 is

$$
\begin{aligned}
\dot{V}_1 &= z_1 \dot{z}_1 - \tilde{\theta}_1^T \Gamma^{-1} \dot{\hat{\theta}} \\
&= z_1(z_2 + \alpha_1 + \phi_1^T \hat{\theta} + \psi_1) - \tilde{\theta}^T (\Gamma^{-1} \dot{\hat{\theta}}_1 - \phi_1 z_1) \\
&= -c_1 z_1^2 + \tilde{\theta}^T (\tau_1 - \Gamma^{-1} \dot{\hat{\theta}}) + z_1 z_2 \qquad (2.7) \\
\tau_1 &= \phi_1 z_1, \qquad (2.8)
\end{aligned}
$$

where τ_1 is the first tuning function to overcome the over-parametrization problem.

Step 2. With the second equation of (2.1) and (2.3), the z_2 dynamics can be derived

$$
\begin{aligned}
\dot{z}_2 &= \dot{x}_2 - \dot{\alpha}_1 - \ddot{x}_r \\
&= x_3 + \phi_2^T \theta + \psi_2 - \frac{\partial \alpha_1}{\partial x_1}(x_2 + \phi_1^T \theta + \psi_1) - \frac{\partial \alpha_1}{\partial \hat{\theta}} \dot{\hat{\theta}} - \frac{\partial \alpha_1}{\partial x_r} \dot{x}_r \\
&= z_3 + \alpha_2 + \psi_2 - \frac{\partial \alpha_1}{\partial x_1}(x_2 + \psi_1) + (\phi_2 - \frac{\partial \alpha_1}{\partial x_1}\phi_1)^T \theta \\
&\quad - \frac{\partial \alpha_1}{\partial \hat{\theta}} \dot{\hat{\theta}} - \frac{\partial \alpha_1}{\partial x_r} \dot{x}_r. \qquad (2.9)
\end{aligned}
$$

Our task in this step is to stabilize the (z_1, z_2)-system (2.4) and (2.9). The Lyapunov function V_2 is chosen as

$$
V_2 = V_1 + \frac{1}{2} z_2^2. \qquad (2.10)
$$

Now we select

$$
\begin{aligned}
\alpha_2 &= -z_1 - c_2 z_2 - \psi_2 + \frac{\partial \alpha_1}{\partial x_1}(x_2 + \psi_1) - \hat{\theta}^T (\phi_2 - \frac{\partial \alpha_1}{\partial x_1}\phi_1) \\
&\quad + \frac{\partial \alpha_1}{\partial \hat{\theta}} \Gamma \tau_2 + \frac{\partial \alpha_1}{\partial x_r} \dot{x}_r \qquad (2.11) \\
\tau_2 &= \tau_1 + (\phi_2 - \frac{\partial \alpha_1}{\partial x_1}\phi_1) z_2, \qquad (2.12)
\end{aligned}
$$

where c_2 is a positive constant and τ_2 is the second tuning function. The resulting derivative of V_2 is

$$
\begin{aligned}
\dot{V}_2 &= -c_1 z_1^2 + z_2 \Big(z_3 + \alpha_2 + z_1 + \psi_2 + \hat{\theta}^T (\phi_2 - \frac{\partial \alpha_1}{\partial x_1}\phi_1) - \frac{\partial \alpha_1}{\partial x_1}(x_2 + \psi_1) \\
&\quad - \frac{\partial \alpha_1}{\partial x_r} \dot{x}_r - \frac{\partial \alpha_1}{\partial \hat{\theta}} \dot{\hat{\theta}} \Big) + \tilde{\theta}^T \Big(\tau_1 + (\phi_2 - \frac{\partial \alpha_1}{\partial x_1}\phi_1) z_2 \Big) - \Gamma^{-1}\dot{\hat{\theta}} \Big) \\
&= -c_1 z_1^2 - c_2 z_2^2 + z_2 z_3 + z_2 \frac{\partial \alpha_1}{\partial \hat{\theta}}(\Gamma \tau_2 - \dot{\hat{\theta}}) + \tilde{\theta}^T(\tau_2 - \Gamma^{-1}\dot{\hat{\theta}}). \qquad (2.13)
\end{aligned}
$$

Step 3. Proceeding to the third equation in (2.1), we obtain

$$
\begin{aligned}
\dot{z}_3 &= z_4 + \alpha_3 + \psi_3 - \frac{\partial \alpha_2}{\partial x_1}(x_2 + \psi_1) - \frac{\partial \alpha_2}{\partial x_2}(x_3 + \psi_2) - \frac{\partial \alpha_2}{\partial x_r} \dot{x}_r - \frac{\partial \alpha_2}{\partial \dot{x}_r} \ddot{x}_r \\
&\quad + (\phi_3 - \frac{\partial \alpha_2}{\partial x_1}\phi_1 - \frac{\partial \alpha_2}{\partial x_2}\phi_2)^T \theta - \frac{\partial \alpha_2}{\partial \hat{\theta}} \dot{\hat{\theta}}. \qquad (2.14)
\end{aligned}
$$

Now we select

$$
\begin{aligned}
\alpha_3 &= -z_2 - c_3 z_3 - \psi_3 + \frac{\partial \alpha_2}{\partial x_1}(x_2 + \psi_1) + \frac{\partial \alpha_2}{\partial x_2}(x_3 + \psi_2) + \frac{\partial \alpha_2}{\partial x_r}\dot{x}_r \\
&\quad + \frac{\partial \alpha_2}{\partial \dot{x}_r}\ddot{x}_r + (\frac{\partial \alpha_1}{\partial \hat{\theta}}\Gamma z_2 - \hat{\theta}^T)(\phi_3 - \frac{\partial \alpha_2}{\partial x_1}\phi_1 - \frac{\partial \alpha_2}{\partial x_2}\phi_2) + \frac{\partial \alpha_2}{\partial \hat{\theta}}\Gamma \tau_3
\end{aligned}
\tag{2.15}
$$

$$
\tau_3 = \tau_2 + (\phi_3 - \frac{\partial \alpha_2}{\partial x_1}\phi_1 - \frac{\partial \alpha_2}{\partial x_2}\phi_2)z_3,
\tag{2.16}
$$

where c_3 is a positive constant. The Lyapunov function is defined as

$$
V_3 = V_2 + \frac{1}{2}z_3^2.
\tag{2.17}
$$

The derivative of the Lapunov function V_3 is

$$
\begin{aligned}
\dot{V}_3 &= -\sum_{i=1}^{3} c_i z_i^2 + z_3 z_4 + z_2 \frac{\partial \alpha_1}{\partial \hat{\theta}}(\Gamma \tau_2 - \dot{\hat{\theta}}) + \tilde{\theta}^T(\tau_2 - \Gamma^{-1}\dot{\hat{\theta}}) \\
&\quad + z_3(\frac{\partial \alpha_1}{\partial \hat{\theta}}\Gamma z_2 + \tilde{\theta}^T)(\phi_3 - \frac{\partial \alpha_2}{\partial x_1}\phi_1 - \frac{\partial \alpha_2}{\partial x_2}\phi_2) + z_3 \frac{\partial \alpha_2}{\partial \hat{\theta}}(\Gamma \tau_3 - \dot{\hat{\theta}}) \\
&= -\sum_{i=1}^{3} c_i z_i^2 + z_3 z_4 + z_2 \frac{\partial \alpha_1}{\partial \hat{\theta}}\left(\Gamma \tau_2 + \Gamma z_3(\phi_3 - \frac{\partial \alpha_2}{\partial x_1}\phi_1 - \frac{\partial \alpha_2}{\partial x_2}\phi_2) - \dot{\hat{\theta}}\right) \\
&\quad + \tilde{\theta}^T\left(\tau_2 + z_3(\phi_3 - \frac{\partial \alpha_2}{\partial x_1}\phi_1 - \frac{\partial \alpha_2}{\partial x_2}\phi_2) - \Gamma^{-1}\dot{\hat{\theta}}\right) + z_3 \frac{\partial \alpha_2}{\partial \hat{\theta}}(\Gamma \tau_3 - \dot{\hat{\theta}}).
\end{aligned}
\tag{2.18}
$$

Note that

$$
\begin{aligned}
\Gamma \tau_2 - \dot{\hat{\theta}} &= \Gamma \tau_2 - \Gamma \tau_3 + \Gamma \tau_3 - \dot{\hat{\theta}} \\
&= -\Gamma z_3(\phi_3 - \frac{\partial \alpha_2}{\partial x_1}\phi_1 - \frac{\partial \alpha_2}{\partial x_2}\phi_2) + (\Gamma \tau_3 - \dot{\hat{\theta}})
\end{aligned}
\tag{2.19}
$$

Then we have

$$
\begin{aligned}
\dot{V}_3 &= -\sum_{i=1}^{3} c_i z_i^2 + z_3 z_4 + \left(z_2 \frac{\partial \alpha_1}{\partial \hat{\theta}} + z_3 \frac{\partial \alpha_2}{\partial \hat{\theta}}\right)\left(\Gamma \tau_3 - \dot{\hat{\theta}}\right) \\
&\quad + \tilde{\theta}^T\left(\tau_3 - \Gamma^{-1}\dot{\hat{\theta}}\right).
\end{aligned}
\tag{2.20}
$$

We can see that the virtual control law α_3 contains the term $\frac{\partial \alpha_1}{\partial \hat{\theta}}\Gamma z_2(\phi_3 - \frac{\partial \alpha_2}{\partial x_1}\phi_1 - \frac{\partial \alpha_2}{\partial x_2}\phi_2)$. This term is an important term, since it is used to cancel the term $z_2 \frac{\partial \alpha_1}{\partial \hat{\theta}}\Gamma(\tau_2 - \tau_3)$ in the derivative $\dot{V}_3$ of the Lyapunov function by using (2.19).

Step i, $(i = 4, \ldots, n)$. Repeating the procedure in a recursive manner, we derive the i-th tracking error for z_i

$$
\begin{aligned}
\dot{z}_i &= z_{i+1} + \alpha_i + \psi_i - \sum_{j=1}^{i-1} \frac{\partial \alpha_{i-1}}{\partial x_j}(x_{j+1} + \psi_j) + \theta^T \left(\phi_i - \sum_{j=1}^{i-1} \frac{\partial \alpha_{i-1}}{\partial x_j}\phi_j\right) \\
&\quad - \frac{\partial \alpha_{i-1}}{\partial \hat{\theta}}\dot{\hat{\theta}} - \sum_{j=1}^{i-1} \frac{\partial \alpha_{i-1}}{\partial x_r^{(j-1)}} x_r^{(j)}.
\end{aligned}
\tag{2.21}
$$

We select the stabilizing function α_i

$$
\begin{aligned}
\alpha_i &= -c_i z_i - z_{i-1} - \psi_i + \sum_{j=1}^{i-1} \frac{\partial \alpha_{i-1}}{\partial x_j}(x_{j+1} + \psi_j) - \hat{\theta}^T \left(\phi_i - \sum_{j=1}^{i-1} \frac{\partial \alpha_{i-1}}{\partial x_j}\phi_j\right) \\
&\quad + \frac{\partial \alpha_{i-1}}{\partial \hat{\theta}}\Gamma\tau_i + \left(\sum_{j=2}^{i-1} z_j \frac{\partial \alpha_{j-1}}{\partial \hat{\theta}}\right)\Gamma\left(\phi_i - \sum_{j=1}^{i-1} \frac{\partial \alpha_{i-1}}{\partial x_j}\phi_j\right) \\
&\quad + \sum_{j=1}^{i-1} \frac{\partial \alpha_{i-1}}{\partial x_r^{(j-1)}} x_r^{(j)}
\end{aligned}
\tag{2.22}
$$

and tuning function

$$
\tau_i = \tau_{i-1} + \left(\phi_i - \sum_{j=1}^{i-1} \frac{\partial \alpha_{i-1}}{\partial x_j}\phi_j\right)z_i,
\tag{2.23}
$$

with the Lyapunov function

$$
V_i = V_{i-1} + \frac{1}{2}z_i^2.
\tag{2.24}
$$

Its derivative is given as

$$
\dot{V}_i = -\sum_{j=1}^{i} c_j z_j^2 + z_i z_{i+1} + \left(\sum_{j=2}^{i} z_j \frac{\partial \alpha_{j-1}}{\partial \hat{\theta}}\right)(\Gamma\tau_i - \dot{\hat{\theta}}) + \tilde{\theta}^T(\tau_i - \Gamma^{-1}\dot{\hat{\theta}}).
\tag{2.25}
$$

In the last step n, the actual control input u appears and is at our disposal. We derive the z_n dynamics as

$$
\begin{aligned}
\dot{z}_n &= bu + \psi_n - \sum_{j=1}^{n-1} \frac{\partial \alpha_{i-1}}{\partial x_j}(x_{j+1} + \psi_j) + \theta^T \left(\phi_n - \sum_{j=1}^{n-1} \frac{\partial \alpha_{i-1}}{\partial x_j}\phi_j\right) \\
&\quad - \frac{\partial \alpha_{n-1}}{\partial \hat{\theta}}\dot{\hat{\theta}} - \sum_{j=1}^{n-1} \frac{\partial \alpha_{i-1}}{\partial x_r^{(j-1)}} x_r^{(j)} - x_r^{(n)}.
\end{aligned}
\tag{2.26}
$$

We are finally in this position to design the control u and the update laws $\dot{\hat{\theta}}$ and $\dot{\hat{p}}$ as

$$u = \hat{p}\bar{u} \tag{2.27}$$

$$\bar{u} = \alpha_n + x_r^{(n)} \tag{2.28}$$

$$\dot{\hat{\theta}} = \Gamma\tau_n \tag{2.29}$$

$$\dot{\hat{p}} = -\gamma\text{sign}(b)\bar{u}z_n, \tag{2.30}$$

where γ is a positive constant and $\hat{p}$ is an estimate of $p = 1/b$. Note that

$$bu = b\hat{p}\bar{u} = \bar{u} - b\tilde{p}\bar{u}, \tag{2.31}$$

where $\tilde{p} = p - \hat{p}$. A block diagram of the backstepping controller design is given in Figure 2.1.

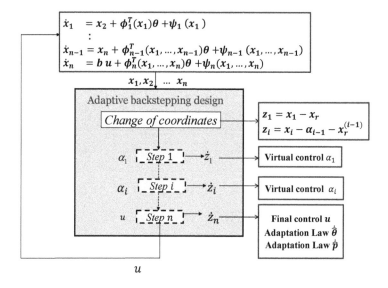

Figure 2.1 Block diagram of the backstepping controller design

2.1.2   Stability Analysis

The main result is summarized in the following theorem.

Theorem 2.1
Consider the closed-loop adaptive system (2.1) under Assumptions 2.1–2.2, the

adaptive controller (2.27), virtual control laws (2.5), (2.11), and (2.22), and updating laws (2.29) and (2.30) guarantee global boundedness of $x(t)$ and $\hat{\theta}, \hat{p}$ and the asymptotic tracking $\lim_{t \to \infty}(x_1 - x_r) = 0$.

Proof 2.1 We choose the Lyapunov function

$$V_n = V_{n-1} + \frac{|b|}{2\gamma}\tilde{p}^2 = \sum_{i=1}^{n}\frac{1}{2}z_i^2 + \frac{1}{2}\tilde{\theta}^T \Gamma^{-1}\tilde{\theta} + \frac{|b|}{2\gamma}\tilde{p}^2, \tag{2.32}$$

where γ is a positive design parameter. Then its derivative is given by

$$\begin{aligned} \dot{V}_n &= -\sum_{i=1}^{n}c_i z_i^2 + \left(\sum_{j=2}^{n}z_j\frac{\partial \alpha_{j-1}}{\partial \hat{\theta}}\right)(\Gamma\tau_n - \dot{\hat{\theta}}) \\ &\quad + \tilde{\theta}^T(\tau_n - \Gamma^{-1}\dot{\hat{\theta}}) - \frac{|b|}{\gamma}\tilde{p}\left(\dot{\hat{p}} + \gamma\text{sign}(b)\bar{u}z_n\right) \\ &= -\sum_{i=1}^{n}c_i z_i^2 \leq 0, \tag{2.33} \end{aligned}$$

From the Lasalle's Theorem in Appendix, this Lyapunov function provides the proof of uniform stability, such that $z_1, z_2, \ldots, z_n, \hat{\theta}, \hat{p}$ are bounded and $z_i \to 0, i = 1, \ldots, n$. This further implies that $\lim_{t \to \infty}(x_1 - x_r) = 0$. Since $x_1 = z_1 + x_r$, x_1 is also bounded from the boundedness of z_1 and x_r. The boundedness of x_2 follows from boundedness of $\dot{x}_r$ and α_1 in (2.5) and the fact that $x_2 = z_2 + \alpha_1 + \dot{x}_r$. Similarly, the boundedness of x_i ($i = 3, \ldots, n$) can be ensured from the boundedness of $x_r^{(i-1)}$ and α_i in (2.22) and the fact that $x_i = z_i + \alpha_{i-1} + x_r^{(i-1)}$. Combining this with (2.27) we conclude that the control $u(t)$ is also bounded. Therefore, boundedness of all signals and asymptotic tracking are ensured as formally stated in the Theorem.

The controller designed in this section achieves the goals of stabilization and tracking. The proof of these properties is a direct consequence of the recursive procedure, because a Lyapunov function is constructed for the entire system including the parameter estimates. By using tuning functions, only one update law is used to estimate unknown parameter θ. This avoids the over-parametrization problem and reduces the dynamic order of the controller to its minimum.

2.2 Adaptive Output Feedback Control

It is noted that the state-feedback controllers presented in the previous session are obtained under the assumption that the full states of the system are measurable. However, for many realistic problems, only the plant output is available for measurement. To address this problem, output-feedback based adaptive backstepping controller design scheme is introduced in this session, where state observers are designed to estimate unmeasurable states and the proposed controller can achieve system stabilization and desired tracking performance.

A class of nonlinear uncertain systems is described in the following form, whose nonlinearities depend only on the output y.

$$
\begin{aligned}
\dot{x}_1 &= x_2 + \phi_1^T(y)\theta + \psi_1(y) \\
&\vdots \\
\dot{x}_{\rho-1} &= x_\rho + \phi_{\rho-1}^T(y)\theta + \psi_{\rho-1}(y) \\
\dot{x}_\rho &= x_{\rho+1} + \phi_\rho^T(y)\theta + \psi_\rho(y) + b_m u \\
&\vdots \\
\dot{x}_{n-1} &= x_n + \phi_{n-1}^T(y)\theta + \psi_{n-1}(y) + b_1 u \\
\dot{x}_n &= \phi_n^T(y)\theta + \psi_n(y) + b_0 u \\
y &= x_1,
\end{aligned}
\tag{2.34}
$$

where $x_1, \ldots, x_n$, y and u are system states, output and input, the vector $\theta \in \Re^r$ is constant and unknown, $\phi_i(y) \in \Re^r$, $i = 1, \ldots, n$ are known nonlinear functions, and $b_m, \ldots, b_0$ are unknown constants.

For the development of control laws, the following assumptions are made.

Assumption 2.3 *The sign of b_m is known.*

Assumption 2.4 *The relative degree $\rho = n - m$ is known and the system is minimum phase.*

Assumption 2.5 *The reference signal y_r and its ρ-th order derivatives are piecewise continuous, known and bounded.*

Our problem is to globally stabilize the system (2.34) and also to achieve the asymptotic tracking of y_r by y.

2.2.1 State Estimation Filters

In order to design the desired adaptive output feedback control law, we rewrite the system (2.34) in the following form

$$
\dot{x} = Ax + \Phi(y)\theta + \Psi(y) + \begin{bmatrix} 0 \\ b \end{bmatrix} u
\tag{2.35}
$$

where

$$A = \begin{bmatrix} 0 & 1 & 0 & \cdots & 0 \\ 0 & 0 & 1 & \cdots & 0 \\ \vdots & \vdots & \vdots & \ddots & \vdots \\ 0 & 0 & 0 & \cdots & 1 \\ 0 & 0 & 0 & \cdots & 0 \end{bmatrix}, \Psi(y) = \begin{bmatrix} \psi_1(y) \\ \vdots \\ \psi_n(y) \end{bmatrix} \tag{2.36}$$

$$\Phi(y) = \begin{bmatrix} \phi_1^T(y) \\ \vdots \\ \phi_n^T(y) \end{bmatrix}, b = \begin{bmatrix} b_m \\ \vdots \\ b_0 \end{bmatrix}. \tag{2.37}$$

Note that only output y is measured and $(x_2, \ldots, x_n)$ are unavailable. We need to design filters to estimate x and generate some signals for controller design. These filters are summarized as

$$\dot{\xi} = A_0\xi + ky + \Psi(y) \tag{2.38}$$
$$\dot{\Xi}^T = A_0\Xi^T + \Phi(y) \tag{2.39}$$
$$\dot{\lambda} = A_0\lambda + e_n u \tag{2.40}$$
$$v_i = A_0^i\lambda, \quad i = 0, 1, \ldots, m, \tag{2.41}$$

where $k = [k_1, \ldots, k_n]^T$ such that all eigenvalues of $A_0 = A - ke_1^T$ are at some desired stable locations, or A_0 is Hurwitz. The state estimates are given by

$$\hat{x}(t) = \xi + \Xi^T\theta + \sum_{i=0}^{m} b_i v_i. \tag{2.42}$$

Note that $\hat{x}$ is unavailable due to the unknown parameters θ and b, so the estimate $\hat{x}$ cannot be used in the controller design. Instead, it will be used for stability analysis. The derivative of $\hat{x}$ is given as

$$\begin{aligned} \dot{\hat{x}}(t) &= \dot{\xi} + \dot{\Xi}^T\theta + \sum_{i=0}^{m} b_i \dot{v}_i \\ &= A_0\xi + ky + \Psi(y) + (A_0\Xi^T + \Phi(y))\theta + \sum_{i=0}^{m} b_i A_0^i(A_0\lambda + e_n u) \\ &= A_0(\xi + \Xi^T\theta + \sum_{i=0}^{m} b_i v_i) + ky + \Phi(y)\theta + \Psi(y) + \begin{bmatrix} 0 \\ b \end{bmatrix} u \\ &= A_0\hat{x} + ky + \Phi(y)\theta + \Psi(y) + \begin{bmatrix} 0 \\ b \end{bmatrix} u. \end{aligned} \tag{2.43}$$

It can be shown that the state estimation error

$$\epsilon = x(t) - \hat{x}(t) \tag{2.44}$$

satisfies

$$
\begin{aligned}
\dot{\epsilon} &= \dot{x}(t) - \dot{\hat{x}}(t) \\
&= Ax - ky - A_0\hat{x} \\
&= (A_0 + ke_1^T)x - ky - A_0\hat{x} \\
&= A_0\epsilon.
\end{aligned}
\tag{2.45}
$$

Suppose $P \in R^{n \times n}$ is a positive definite matrix, satisfying $PA_0 + A_0^T P \leq -I$ and let

$$
V_\epsilon = \epsilon^T P \epsilon.
\tag{2.46}
$$

It can be shown that

$$
\begin{aligned}
\dot{V}_\epsilon &= \epsilon^T (PA_0 + A_0^T P)\epsilon \\
&\leq -\epsilon^T \epsilon.
\end{aligned}
\tag{2.47}
$$

This Lyapunov function guarantees that $\epsilon \to 0$, which implies $\hat{x}(t) \to x(t)$.

Note that the backstepping design starts with its output y, which is the only available system state allowed to appear in the control law. The dynamic equation of y is expressed as

$$
\begin{aligned}
\dot{y} &= x_2 + \phi_1^T(y)\theta(t) + \psi_1(y) \\
&= b_m v_{m,2} + \xi_2 + \psi_1(y) + \bar{\omega}^T \Theta + \epsilon_2.
\end{aligned}
\tag{2.48}
$$

where

$$
\begin{aligned}
\Theta &= [b_m, \ldots, b_0, \theta^T]^T \tag{2.49} \\
\omega &= [v_{m,2}, v_{m-1,2}, \ldots, v_{0,2}, \Xi_2 + \phi_1^T]^T \tag{2.50} \\
\bar{\omega} &= [0, v_{m-1,2}, \ldots, v_{0,2}, \Xi_2 + \phi_1^T]^T. \tag{2.51}
\end{aligned}
$$

In above equations, $\epsilon_2, v_{i,2}, \xi_2$ and Ξ_2 denote the second entries of ϵ, v_i, ξ, and Ξ, respectively, and y, v_i, ξ, and Ξ are all available signals.

Combining system (2.48) with our filters (2.38)–(2.41), system (2.34) is represented as

$$
\begin{aligned}
\dot{y} &= b_m v_{m,2} + \xi_2 + \psi_1(y) + \bar{\omega}^T \Theta + \epsilon_2 \tag{2.52} \\
\dot{v}_{m,i} &= v_{m,i+1} - k_i v_{m,1}, \qquad i = 2, 3, \ldots, \rho - 1 \tag{2.53} \\
\dot{v}_{m,\rho} &= v_{m,\rho+1} - k_\rho v_{m,1} + u. \tag{2.54}
\end{aligned}
$$

System (2.52)–(2.54) will be our design system, whose states $y, v_{m,2}, \ldots, v_{m,\rho}$ are available. Our task at this stage is to globally stabilize the system and also to achieve the asymptotic tracking of y_r by y.

2.2.2 Design of Adaptive Controllers and Stability Analysis

In this section, we present the adaptive control design using the backstepping technique with tuning functions in ρ steps. Firstly, we take the change of coordinates

$$z_1 = y - y_r \tag{2.55}$$
$$z_i = v_{m,i} - \alpha_{i-1} - \hat{\varrho} y_r^{(i-1)}, \quad i = 2, 3, \ldots, \rho, \tag{2.56}$$

where $\hat{\varrho}$ is an estimate of $\varrho = 1/b_m$ and α_{i-1} is the virtual control at each step $i = 2, 3, \ldots, \rho$.

• *Step* 1: Starting with the equation for the tracking error z_1, we obtain, from (2.52) and (2.55), that

$$\dot{z}_1 = b_m v_{m,2} + \xi_2 + \psi_1(y) + \bar{\omega}^T \Theta + \epsilon_2 - \dot{y}_r. \tag{2.57}$$

By substituting (2.56) for $i = 2$ into (2.57) and using $\tilde{\varrho} = \frac{1}{b_m} - \frac{1}{\hat{b}_m}$, we get

$$\dot{z}_1 = b_m \alpha_1 + \xi_2 + \psi_1(y) + \bar{\omega}^T \Theta + \epsilon_2 - b_m \tilde{\varrho} \dot{y}_r + b_m z_2. \tag{2.58}$$

By considering $v_{m,2}$ as the first virtual control, we select a virtual control law α_1 as

$$\alpha_1 = \hat{\varrho} \bar{\alpha}_1 \tag{2.59}$$
$$\bar{\alpha}_1 = -c_1 z_1 - d_1 z_1 - \xi_2 - \psi_1(y) - \bar{\omega}^T \hat{\Theta}, \tag{2.60}$$

where c_1 and d_1 are positive design parameters, and $\hat{\Theta}$ is the estimate of Θ. From (2.58) and (2.59), we have

$$
\begin{aligned}
\dot{z}_1 &= -c_1 z_1 - d_1 z_1 + \epsilon_2 + \bar{\omega}^T \tilde{\Theta} - b_m(\dot{y}_r + \bar{\alpha}_1)\tilde{\varrho} + b_m z_2 \\
&= -(c_1 + d_1)z_1 + \epsilon_2 + (\omega - \hat{\varrho}(\dot{y}_r + \bar{\alpha}_1)e_1)^T \tilde{\Theta} - b_m(\dot{y}_r + \bar{\alpha}_1)\tilde{\varrho} + \hat{b}_m z_2,
\end{aligned}
\tag{2.61}
$$

where $\tilde{\Theta} = \Theta - \hat{\Theta}$. Note that

$$b_m \alpha_1 = b_m \hat{\varrho} \bar{\alpha}_1 = \bar{\alpha}_1 - b_m \tilde{\varrho} \bar{\alpha}_1 \tag{2.62}$$
$$
\begin{aligned}
\bar{\omega}^T \tilde{\Theta} + b_m z_2 &= \bar{\omega}^T \tilde{\Theta} + \tilde{b}_m z_2 + \hat{b}_m z_2 \\
&= \bar{\omega}^T \tilde{\Theta} + (v_{m,2} - \hat{\varrho} \dot{y}_r - \alpha_1)e_1^T \tilde{\Theta} + \hat{b}_m z_2 \\
&= (\omega - \hat{\varrho}(\dot{y}_r + \bar{\alpha}_1)e_1)^T \tilde{\Theta} + \hat{b}_m z_2,
\end{aligned}
\tag{2.63}
$$

Define the Lyapunov function V_1 as

$$V_1 = \frac{1}{2}z_1^2 + \frac{1}{2}\tilde{\Theta}^T \Gamma^{-1} \tilde{\Theta} + \frac{|b_m|}{2\gamma}\tilde{\varrho}^2 + \frac{1}{2d_1}\epsilon^T P \epsilon, \tag{2.64}$$

where Γ is a positive definite design matrix, γ is a positive design parameter, and P is a definite positive matrix such that $PA_0 + A_0^T P = -I$, $P = P^T > 0$. We examine the derivative of V_1

$$
\begin{aligned}
\dot{V}_1 \;\le\;& z_1\dot{z}_1 - \tilde{\Theta}^T\Gamma^{-1}\dot{\hat{\Theta}} - \frac{|b_m|}{\gamma}\tilde{\varrho}\dot{\hat{\varrho}} - \frac{1}{2d_1}\epsilon^T\epsilon \\
\;\le\;& -c_1 z_1^2 + \hat{b}_m z_1 z_2 - \frac{1}{4d_1}\epsilon^T\epsilon - |b_m|\tilde{\varrho}\frac{1}{\gamma}[\gamma\mathrm{sign}(b_m)(\dot{y}_r + \bar{\alpha}_1)z_1 + \dot{\hat{\varrho}}] \\
& +\tilde{\Theta}^T[(\omega - \hat{\varrho}(\dot{y}_r + \bar{\alpha}_1)e_1)z_1 - \Gamma^{-1}\dot{\hat{\Theta}}] - d_1 z_1^2 + z_1\epsilon_2 - \frac{\|\epsilon\|^2}{4d_1}. \quad (2.65)
\end{aligned}
$$

Now we choose

$$
\dot{\hat{\varrho}} \;=\; -\gamma\mathrm{sign}(b_m)(\dot{y}_r + \bar{\alpha}_1)z_1. \qquad (2.66)
$$

Define

$$
\tau_1 \;=\; (\omega - \hat{\varrho}(\dot{y}_r + \bar{\alpha}_1)e_1)z_1, \qquad (2.67)
$$

and τ_1 is called the first tuning function. Then the following can be derived by using Young's inequality $ab \le d_1 a^2 + \frac{1}{4d_1}b^2$, update law (2.66) and (2.67).

$$
\dot{V}_1 \;\le\; -c_1 z_1^2 + \hat{b}_m z_1 z_2 - \frac{1}{4d_1}\epsilon^T\epsilon + \tilde{\Theta}^T(\tau_1 - \Gamma^{-1}\dot{\hat{\Theta}}). \qquad (2.68)
$$

If $z_2 = 0$, we would choose $\dot{\hat{\Theta}} = \Gamma\tau_1$ and the derivative of V_1 would be

$$
\dot{V}_1 \;\le\; -c_1 z_1^2 - \frac{1}{4d_1}\epsilon^T\epsilon \le -c_1 z_1^2, \qquad (2.69)
$$

which implies that z_1 converges to zero asymptotically. Since $z_2 \ne 0$, we do not use $\dot{\hat{\Theta}} = \Gamma\tau_1$ as an update law for Θ at this step to avoid over-parametrization problem, because Θ will also appear in the following steps.

• *Step* 2: We derive the error dynamics z_2

$$
\begin{aligned}
\dot{z}_2 \;=\;& \dot{v}_{m,2} - \dot{\alpha}_1 - \dot{\hat{\varrho}}\dot{y}_r - \hat{\varrho}\ddot{y}_r \\
\;=\;& v_{m,3} - k_2 v_{m,1} - \frac{\partial\alpha_1}{\partial y}(b_m v_{m,2} + \xi_2 + \psi_1 + \bar{\omega}^T\Theta + \epsilon_2) - \frac{\partial\alpha_1}{\partial y_r}\dot{y}_r \\
& -\sum_{j=1}^{m+i-1}\frac{\partial\alpha_1}{\partial\lambda_j}(-k_j\lambda_1 + \lambda_{j+1}) - \frac{\partial\alpha_1}{\partial\xi}(A_0\xi + ky + \Psi(y)) \\
& -\frac{\partial\alpha_1}{\partial\Xi}(A_0\Xi^T + \Phi(y)) - \frac{\partial\alpha_1}{\partial\hat{\Theta}}\dot{\hat{\Theta}} - \frac{\partial\alpha_1}{\partial\hat{\varrho}}\dot{\hat{\varrho}} - \dot{\hat{\varrho}}\dot{y}_r - \hat{\varrho}\ddot{y}_r \\
\;=\;& v_{m,3} - \hat{\varrho}\ddot{y}_r - \beta_2 - \frac{\partial\alpha_1}{\partial y}(\omega^T\tilde{\Theta} + \epsilon_2)) - \frac{\partial\alpha_1}{\partial\hat{\Theta}}\dot{\hat{\Theta}}, \qquad (2.70)
\end{aligned}
$$

where

$$
\begin{aligned}
\beta_2 \;=\; & \frac{\partial \alpha_1}{\partial y}(\xi_2 + \psi_1 + \omega^T \hat{\Theta}) + k_2 v_{m,1} + \frac{\partial \alpha_1}{\partial y_r}\dot{y}_r + \left(\dot{y}_r + \frac{\partial \alpha_1}{\partial \hat{\varrho}}\right)\dot{\hat{\varrho}} \\
& + \sum_{j=1}^{m+i-1} \frac{\partial \alpha_1}{\partial \lambda_j}(-k_j \lambda_1 + \lambda_{j+1}) + \frac{\partial \alpha_1}{\partial \xi}(A_0 \xi + ky + \Psi(y)) \\
& + \frac{\partial \alpha_1}{\partial \Xi^T}(A_0 \Xi^T + \Phi(y))
\end{aligned}
\tag{2.71}
$$

By considering $v_{m,3}$ as virtual control input and using $z_3 = v_{m,3} - \alpha_2 - \hat{\varrho}\ddot{y}_r$, we have

$$
\dot{z}_2 \;=\; z_3 + \alpha_2 - \beta_2 - \frac{\partial \alpha_1}{\partial y}(\omega^T \tilde{\Theta} + \epsilon_2) - \frac{\partial \alpha_1}{\partial \hat{\Theta}}\dot{\hat{\Theta}}.
\tag{2.72}
$$

Consider the Lyapunov function

$$
V_2 = V_1 + \frac{1}{2}z_2^2 + \frac{1}{2d_2}\epsilon^T P \epsilon.
\tag{2.73}
$$

We choose the second virtual control law α_2 and tuning function as

$$
\alpha_2 \;=\; -\hat{b}_m z_1 - \left(c_2 + d_2\Big(\frac{\partial \alpha_1}{\partial y}\Big)^2\right)z_2 + \beta_2 + \frac{\partial \alpha_1}{\partial \hat{\Theta}}\Gamma \tau_2
\tag{2.74}
$$

$$
\tau_2 \;=\; \tau_1 - \frac{\partial \alpha_1}{\partial y}\omega z_2.
\tag{2.75}
$$

Then

$$
\begin{aligned}
\dot{V}_2 \;=\; & \dot{V}_1 + z_2 \dot{z}_2 - \frac{1}{2d_2}\epsilon^T \epsilon \\
\leq\; & -c_1 z_1^2 + \hat{b}_m z_1 z_2 + z_2\left(z_3 + \alpha_2 - \beta_2 - \frac{\partial \alpha_1}{\partial y}(\omega^T \tilde{\Theta} + \epsilon_2) - \frac{\partial \alpha_1}{\partial \hat{\Theta}}\dot{\hat{\Theta}}\right) \\
& - \frac{1}{2d_2}\epsilon^T \epsilon - \frac{1}{4d_1}\epsilon^T \epsilon + \tilde{\Theta}^T(\tau_1 - \Gamma^{-1}\dot{\hat{\Theta}}) \\
=\; & -c_1 z_1^2 - c_2 z_2^2 + z_2 z_3 - d_2\Big(\frac{\partial \alpha_1}{\partial y}\Big)^2 z_2^2 - \frac{\partial \alpha_1}{\partial y}\epsilon_2 z_2 - \frac{1}{4d_2}\epsilon^T \epsilon \\
& - \frac{1}{4d_2}\epsilon^T \epsilon - \frac{1}{4d_1}\epsilon^T \epsilon + \tilde{\Theta}^T\Big(\tau_1 - \frac{\partial \alpha_1}{\partial y}\omega z_2 - \Gamma^{-1}\dot{\hat{\Theta}}\Big) + \frac{\partial \alpha_1}{\partial \hat{\Theta}}(\Gamma \tau_2 - \dot{\hat{\Theta}}) \\
\leq\; & -\sum_{i=1}^{2}\left(c_i z_i^2 + \frac{1}{4d_i}\epsilon^T \epsilon\right) + z_2 z_3 + \tilde{\Theta}^T(\tau_2 - \Gamma^{-1}\dot{\hat{\Theta}}) + \frac{\partial \alpha_1}{\partial \hat{\Theta}}(\Gamma \tau_2 - \dot{\hat{\Theta}}).
\end{aligned}
\tag{2.76}
$$

Following similar arguments as before, we would choose $\dot{\hat{\Theta}} = \Gamma \tau_2$, as this would result in $\dot{V}_2 \leq -c_1 z_1^2 - c_2 z_2^2$ if $z_3 = 0$. But $z_3 \neq 0$ and thus we do not use it as an update law for Θ to overcome the over-parametrization problem.

● *Step i ($i = 3, \ldots, \rho$)*: Choose virtual control laws

$$\alpha_i = -z_{i-1} - \left[c_i + d_i \left(\frac{\partial \alpha_{i-1}}{\partial y}\right)^2\right] z_i + \beta_i + \frac{\partial \alpha_{i-1}}{\partial \hat{\Theta}} \Gamma \tau_i$$

$$-\left(\sum_{k=2}^{i-1} z_k \frac{\partial \alpha_{k-1}}{\partial \hat{\Theta}}\right) \Gamma \frac{\partial \alpha_{i-1}}{\partial y} \omega, \quad i = 3, \ldots, \rho, \tag{2.77}$$

where c_i are positive design parameters and

$$\tau_i = \tau_{i-1} - \frac{\partial \alpha_{i-1}}{\partial y} \omega z_i \tag{2.78}$$

$$\beta_i = \frac{\partial \alpha_{i-1}}{\partial y}(\xi_2 + \psi_1 + \omega^T \hat{\Theta}) + k_i v_{m,1} + \sum_{j=1}^{i-1} \frac{\partial \alpha_{i-1}}{\partial y_r^{(j-1)}} y_r^{(j)}$$

$$+ \left(y_r^{(i-1)} + \frac{\partial \alpha_{i-1}}{\partial \hat{\varrho}}\right) \dot{\hat{\varrho}} + \sum_{j=1}^{m+i-1} \frac{\partial \alpha_{i-1}}{\partial \lambda_j}(-k_j \lambda_1 + \lambda_{j+1})$$

$$+ \frac{\partial \alpha_{i-1}}{\partial \xi}(A_0 \xi + ky + \Psi(y)) + \frac{\partial \alpha_{i-1}}{\partial \Xi^T}(A_0 \Xi^T + \Phi(y)). \tag{2.79}$$

In the last step ρ, the adaptive controller and parameter update law are given by

$$u = \alpha_\rho - v_{m,\rho+1} + \hat{\varrho} y_r^{(\rho)} \tag{2.80}$$

$$\dot{\hat{\Theta}} = \Gamma \tau_\rho \tag{2.81}$$

We define the final Lyapunov function V_ρ as

$$V_\rho = \sum_{i=1}^{\rho} \frac{1}{2} z_i^2 + \frac{1}{2} \tilde{\Theta}^T \Gamma^{-1} \tilde{\Theta} + \frac{|b_m|}{2\gamma} \tilde{\varrho}^2 + \sum_{i=1}^{\rho} \frac{1}{2d_i} \epsilon^T P \epsilon. \tag{2.82}$$

Note that

$$\Gamma \tau_{i-1} - \dot{\hat{\Theta}} = \Gamma \tau_{i-1} - \Gamma \tau_i + \Gamma \tau_i - \dot{\hat{\Theta}}$$

$$= \Gamma \frac{\partial \alpha_{i-1}}{\partial y} \omega z_i + (\Gamma \tau_i - \dot{\hat{\Theta}}). \tag{2.83}$$

From (2.77)–(2.81), the derivative of the last Lyapunov function satisfies

$$\dot{V}_\rho = \sum_{i=1}^{\rho} z_i \dot{z}_i - \tilde{\Theta}^T \Gamma^{-1} \dot{\hat{\Theta}} - \frac{|b_m|}{\gamma} \tilde{\varrho} \dot{\hat{\varrho}} - \sum_{i=1}^{\rho} \frac{1}{2d_i} \epsilon^T \epsilon$$

$$\leq -\sum_{i=1}^{\rho} c_i z_i^2 - \sum_{i=1}^{\rho} \frac{1}{4d_i} \epsilon^T \epsilon - \tilde{\Theta}^T \Gamma^{-1}(\dot{\hat{\Theta}} - \Gamma \tau_\rho)$$

$$+ \left(\sum_{k=2}^{\rho} z_k \frac{\partial \alpha_{k-1}}{\partial \hat{\Theta}}\right) (\Gamma \tau_\rho - \dot{\hat{\Theta}})$$

$$= -\sum_{i=1}^{\rho} c_i z_i^2 - \sum_{i=1}^{\rho} \frac{1}{4d_i} \epsilon^T \epsilon. \tag{2.84}$$

We have the following stability and performance results based on the designed backstepping controller.

Theorem 2.2

Consider the system (2.34) under assumptions 2.3–2.5, consisting of the parameter estimators given by (2.66) and (2.81), adaptive controllers designed using (2.80) with virtual control laws (2.59), (2.74), and (2.77), the filters (2.38), (2.39), and (2.40). The system is stable in the sense that all signals in the closed loop system are globally uniformly bounded. Furthermore,

- *The asymptotic tracking performance is achieved, i.e.*

$$\lim_{t \to \infty} [y(t) - y_r(t)] = 0. \tag{2.85}$$

- *The transient tracking error performance is given by*

$$
\begin{aligned}
\| y(t) - y_r(t) \|_2 \leq \ & \frac{1}{\sqrt{c_1}} \Big(\frac{1}{2} \tilde{\Theta}(0)^T \Gamma^{-1} \tilde{\Theta}(0) + \frac{|b_m|}{2\gamma} \tilde{\varrho}(0)^2 \\
& + \frac{1}{2d_0} \| \epsilon(0) \|_P^2 \Big)^{1/2},
\end{aligned} \tag{2.86}
$$

with $z_i(0) = 0, i = 1, \ldots, \rho,\ d_0 = \left(\sum_{i=1}^{\rho} \frac{1}{d_i} \right)^{-1}$ and $\| \epsilon(0) \|_P^2 = \epsilon(0)^T P \epsilon(0)$.

Proof: Due to the piecewise continuity of $y_r(t), \ldots, y_r^{(\rho)}(t)$ and the smoothness of the control law, the parameter updating laws and the filters, the solution of the closed-loop adaptive system exists and is unique. From (2.84), it can be shown that V_ρ is uniformly bounded. Thus $z_i, \hat{\Theta}, \hat{\varrho},$ and ϵ are bounded. Since z_1 and y_r are bounded, y is also bounded. Then from (2.38) and (2.39), we conclude that ξ and Ξ are bounded as A_0 is Hurwitz. From (2.40) and Assumption 2.4, we have that $\lambda_1, \ldots, \lambda_{m+1}$ are bounded. From the coordinate change (2.56), it gives

$$
\begin{aligned}
v_{m,i} = \ & z_i + \hat{\varrho} y_r^{(i-1)} + \alpha_{i-1}\big(y, \xi, \Xi, \hat{\Theta}, \hat{\varrho}, \bar{\lambda}_{m+i-1}, \bar{y}_r^{(i-2)}\big) \\
& i = 2, 3, \ldots, \rho,
\end{aligned} \tag{2.87}
$$

where $\bar{\lambda}_k = [\lambda_1, \ldots, \lambda_k]^T, \bar{y}_r^{(k)} = [y_r, \ldots, y_r^{(k)}]^T$. For $i = 2$, from the boundedness of $\lambda_{m+1}, z_2, y, \xi, \Xi, \hat{\Theta}, \hat{\varrho}, y_r$ and $\dot{y}_r$, it proves that $v_{m,2}$ is bounded. From (2.41), it follows that λ_{m+2} is bounded. Following the same procedure recursively, we can show that λ is bounded. From (2.42) and the boundedness of $\xi, \Xi, \lambda, \epsilon$, we conclude that x is bounded.

To show the global uniform stability, the boundedness of $m = n - \rho$ dimension states ζ with zero dynamics should be guaranteed. Under a similar transformation as in [53], the states ζ associated with the zero dynamics can be shown to satisfy

$$\dot{\zeta} = A_b \zeta + b_b y + T \Phi(y)\theta + T \Psi(y), \tag{2.88}$$

where $\zeta = Tx$, $b_b \in R^m$, the eigenvalues of the $m \times m$ matrix A_b is given as follows

$$A_b = \begin{bmatrix} -b_{m-1}/b_m & & \\ & & I_{m-1} \\ \vdots & & \\ -b_0/b_m & 0 & \cdots & 0 \end{bmatrix} \qquad (2.89)$$

$$T = [(A_b)^\rho e_1, \ldots, A_b e_1, I_m]. \qquad (2.90)$$

With Assumption 2.4, we have that A_b is Hurwitz. Hence, there exists matrix P such that

$$PA_b + (A_b)^T P = -2I \qquad (2.91)$$

Now we define a Lyapunov function for the zero dynamics of the system as $V_\zeta = \zeta^T P \zeta$. It can be show that

$$\dot{V}_\zeta \leq -\zeta^T \zeta + \| P(b_b y + T\Phi(y)^T \theta + \Psi(y)) \|^2. \qquad (2.92)$$

Because all signals and functions in the second term of (2.92) are bounded, it can be shown that ζ is bounded. Thus all signals in the closed-loop are globally uniformly bounded. By applying the LaSalle-Yoshizawa theorem to (2.84), it further follows that $z(t) \to 0$ as $t \to \infty$, which implies that $\lim_{t\to\infty}[y(t) - y_r(t)] = 0$.

Now we derive the tracking error bound in term of the L_2 norm. As shown in (2.84), the derivative of V_ρ is

$$\dot{V}_\rho \leq -\sum_{i=1}^{\rho} c_i z_i^2 \leq -c_1 z_1^2. \qquad (2.93)$$

Since V_ρ is non-increasing, we have

$$\| z_1 \|_2^2 = \int_0^\infty |z_1(\tau)|^2 d\tau \leq \frac{1}{c_1}(V(0) - V(\infty)) \leq \frac{1}{c_1}V(0). \qquad (2.94)$$

We can set $z_i(0)$ to zero by appropriately initializing the reference trajectory as following

$$y_r(0) = y(0) \qquad (2.95)$$

$$y_r^{(i)}(0) = \frac{1}{\hat{\varrho}(0)}\Big[v_{m,i+1}(0) - \alpha_i\big(y(0), \xi(0), \Xi(0), \hat{\Theta}(0), \hat{\varrho}(0), \bar{\lambda}_{m+i}(0),$$

$$\bar{y}_r^{(i-1)}(0)\big)\Big], \quad i = 1, \ldots, \rho - 1 \qquad (2.96)$$

Thus, by setting $z_i(0) = 0, i = 1, \ldots, n$, we obtain

$$V(0) = \frac{1}{2}\tilde{\Theta}(0)^T \Gamma^{-1} \tilde{\Theta}(0) + \frac{|b_m|}{2\gamma}\tilde{\varrho}(0)^2 + \frac{1}{2d_0} \| \epsilon(0) \|_P^2, \qquad (2.97)$$

which is a decreasing function of γ, η and Γ, independent of c_1. This means that the bound resulting from (2.94) and (2.97) is

$$\| z_1 \|_2 \leq \frac{1}{\sqrt{c_1}} \left(\frac{1}{2} \tilde{\Theta}(0)^T \Gamma^{-1} \tilde{\Theta}(0) + \frac{|b_m|}{2\gamma} \tilde{\varrho}(0)^2 + \frac{1}{2d_0} \| \epsilon(0) \|_P^2 \right)^{1/2} \quad (2.98)$$

△△△

Remark 2.1 The following conclusions can be obtained:
• The transient performance depends on the initial estimate errors $\tilde{\Theta}(0)$, $\tilde{\varrho}(0)$ and the explicit design parameters. The closer the initial estimates $\hat{\Theta}(0)$ and $\hat{\varrho}(0)$ to the true values Θ and ϱ, the better the transient performance.
• The bound for $\| y(t) - y_r(t) \|_2$ is an explicit function of design parameters and thus computable. We can decrease the effects of the initial error estimates on the transient performance by increasing the adaptation gains Γ, γ, d_0, or c_1.

2.3 Robust State Feedback Control

Besides adaptive control, robust control is also an effective method to deal with system uncertainties. In this section, we present a robust state-feedback backstepping method which can handle both unmodeled system dynamics and external disturbances. The considered nonlinear uncertain system is in the following strict-feedback form

$$
\begin{aligned}
\dot{x}_1 &= x_2 + f_1(x_1) + \Delta f_1(x_1) + d_1(t) \\
\dot{x}_i &= x_{i+1} + f_i(x_1, x_2, \ldots, x_i) + \Delta f_2(x_1, x_2, \ldots, x_i) + d_i(t) \\
\dot{x}_n &= u + f_n(x_1, \ldots, x_n) + \Delta f_n(x_1, \ldots, x_n) + d_n(t) \quad (2.99)
\end{aligned}
$$

where $x = [x_1, \ldots, x_n]^T \in \mathbb{R}^n$ and $u \in \mathbb{R}^1$ are the states and control input of the system, respectively. To simplify the notations, let $\bar{x}_i = [x_1, \ldots, x_i]^T \in \mathbb{R}^i (i = 1, \ldots, n-1)$. $f_i(\cdot) : \mathbb{R}^i \rightarrow \mathbb{R}^1 (i = 1, \ldots, n)$, are known nonlinear functions; $\Delta f_i(\cdot) : \mathbb{R}^i \rightarrow \mathbb{R}^1 (i = 1, \ldots, n)$, denote the system uncertainties and $d_i(t)(i = 1, \ldots, n)$, are bounded external time varying disturbances.

The control objective is to design a controller which can make the output $x_1(t)$ track a reference signal $x_r(t)$ while all the other signals are bounded. To this end, the following assumptions are made.

Assumption 2.6 *The uncertain function $\Delta f_i(\bar{x}_i)$ and the external disturbance $d_i(t)$ satisfy that $|\Delta f_i(\bar{x}_i)| \leq \rho_i(\bar{x}_i)$ and $|d_i(t)| \leq D_i$ respectively, where $\rho_i(\bar{x}_i)$ is at least $(n-i+1)$-th order differentiable and D_i is a known positive constant denoting the upper bound of the disturbance.*

Assumption 2.7 *The reference signal $x_r(t)$ is n-th order differentiable.*

2.3.1 Controller Design

In this subsection, a robust controller is designed based on backstepping technique, which involves n recursive steps.

First, the change of coordinates is introduced:

$$\begin{aligned} z_1 &= x_1 - x_r(t) \\ z_i &= x_i - \alpha_{i-1}, \quad i = 2, \ldots, n, \end{aligned} \tag{2.100}$$

where α_i (i=1, 2, ..., n) are the virtual control laws. Following the backstepping design procedure, the design steps are summarized as follows.

- *Step* 1. From (2.99), the derivative of z_1 is

$$\dot{z}_1 = z_2 + \alpha_1 + f_1(x_1) + d_1(t) + \Delta f_1(x_1) - \dot{x}_r, \tag{2.101}$$

The virtual control law α_1 is chosen as

$$\alpha_1 = -c_1 z_1 - f_1(x_1) + \dot{x}_r - \frac{z_1(\rho_1(x_1) + D_1)^2}{2\sigma}, \tag{2.102}$$

where c_1 and σ are positive design constants. Choose the Lyapunov function as

$$V_1 = \frac{1}{2} z_1^2. \tag{2.103}$$

Then its derivative is given as

$$\begin{aligned} \dot{V}_1 &= z_1(z_2 + \alpha_1 + f_1(x_1) + d_1(t) + \Delta f_1(x_1) - \dot{x}r) \\ &\leq z_1(z_2 + \alpha_1 + f_1(x_1) - \dot{x}_r) + |z_1|(\rho_1(x_1) + D_1) \\ &\leq z_1 z_2 + z_1(\alpha_1 + f_1(x_1) - \dot{x}_r) + \frac{z_1^2(\rho_1(x_1) + D_1)^2}{2\sigma} + \frac{\sigma}{2} \\ &\leq z_1 z_2 - c_1 z_1^2 + \frac{\sigma}{2}. \end{aligned} \tag{2.104}$$

where Young's inequality $ab \leq \frac{a^2 b^2}{2\sigma} + \frac{\sigma}{2}$ is used.
- *Step* 2. The derivative of z_2 is

$$\begin{aligned} \dot{z}_2 &= \dot{x}_2 - \dot{\alpha}_1 \\ &= z_3 + \alpha_2 + f_2 + \Delta f_2 + d_2 - \frac{\partial \alpha_1}{\partial x_1}(x_2 + f_1 + \Delta f_1 + d_1) - \ddot{x}_r. \end{aligned} \tag{2.105}$$

We consider the following Lyapunov function

$$V_2 = V_1 + \frac{1}{2} z_2^2. \tag{2.106}$$

The virtual control law is chosen as

$$\begin{aligned} \alpha_2 = &- c_2 z_2 - z_1 - f_2 + \frac{\partial \alpha_1}{\partial x_1}(x_2 + f_1) + \ddot{x}_r \\ &- z_2 \left(\frac{\partial \alpha_1}{\partial x_1} \right)^2 \frac{(\rho_1 + D_1)^2}{2\sigma} - \frac{z_2(\rho_2 + D_2)^2}{2\sigma}. \end{aligned} \tag{2.107}$$

where c_2 is a positive design constant. Thus the resulting of derivative of V_2 is

$$
\dot{V}_2 = \dot{V}_1 + z_2(z_3 + \alpha_2 + f_2 + \Delta f_2 + d_2 - \frac{\partial \alpha_1}{\partial x_1}(x_2 + f_1 + \Delta f_1 + d_1) - \ddot{x}_r)
$$

$$
\leq \dot{V}_1 + z_2(z_3 + \alpha_2 + f_2 - \frac{\partial \alpha_1}{\partial x_1}(x_2 + f_1) - \ddot{x}_r)
$$

$$
+ |z_2|(\rho_2 + D_2) + \left| z_2 \frac{\partial \alpha_1}{\partial x_1} \right| (\rho_1 + D_1)
$$

$$
\leq \dot{V}_1 + z_2 z_3 + z_2 \left(\alpha_2 + f_2 - \frac{\partial \alpha_1}{\partial x_1}(x_2 + f_1) - \ddot{x}_r \right)
$$

$$
+ \frac{1}{2\sigma}(z_2)^2(\rho_2 + D_2)^2 + \frac{1}{2\sigma}(z_2)^2 \left(\frac{\partial \alpha_1}{\partial x_1} \right)^2 (\rho_1 + D_1)^2 + \frac{2\sigma}{2} \qquad (2.108)
$$

Substituting (2.104) and (2.107) into (2.108), we obtain

$$
\dot{V}_2 \leq -c_1 z_1^2 - c_2 z_2^2 + z_2 z_3 + \frac{3}{2}\sigma. \qquad (2.109)
$$

• *Step* 3. The derivative of z_3 is

$$
\dot{z}_3 = z_4 + \alpha_3 + f_3 + \Delta f_3 + d_3 - \dot{\alpha}_2
$$

$$
= z_4 + \alpha_3 + f_3 + \Delta f_3 + d_3 - \frac{\partial \alpha_2}{\partial x_1}(x_2 + f_1 + \Delta f_1 + d_1)
$$

$$
- \frac{\partial \alpha_2}{\partial x_2}(x_3 + f_2 + \Delta f_2 + d_2) - x_r^{(3)}. \qquad (2.110)
$$

The Lyapunov funtion is chosen as

$$
V_3 = V_2 + \frac{1}{2}z_3^2. \qquad (2.111)
$$

The virtual control law is chosen as

$$
\alpha_3 = - c_3 z_3 - z_2 - f_3 + \frac{\partial \alpha_2}{\partial x_1}(f_1 + x_2) + \frac{\partial \alpha_2}{\partial x_2}(f_2 + x_3) + x_r^{(3)}
$$

$$
- \frac{z_3(\rho_3 + D_3)^2}{2\sigma} - z_3 \left(\frac{\partial \alpha_2}{\partial x_1} \right)^2 \frac{(\rho_1 + D_1)^2}{2\sigma} - z_3 \left(\frac{\partial \alpha_2}{\partial x_2} \right)^2 \frac{(\rho_2 + D_2)^2}{2\sigma},
$$

$$
(2.112)
$$

where c_3 is a positive design constant. Thus the resulting of derivative of V_3 is

$$
\begin{aligned}
\dot{V}_3 = {} & \dot{V}_2 + z_3(z_4 + \alpha_3 + f_3 - \frac{\partial \alpha_2}{\partial x_1}(x_2 + f_1) - \frac{\partial \alpha_2}{\partial x_2}(x_3 + f_2) - r^{(3)}) \\
& + z_3(\Delta f_3 + d_3) - z_3 \frac{\partial \alpha_2}{\partial x_1}(\Delta f_1 + d_1) - z_3 \frac{\partial \alpha_2}{\partial x_2}(\Delta f_2 + d_2) \\
\leq {} & \dot{V}_2 + z_3 z_4 + z_3 \left(\alpha_3 + f_3 - \frac{\partial \alpha_2}{\partial x_1}(x_2 + f_1) - \frac{\partial \alpha_2}{\partial x_2}(x_3 + f_2) - x_r^{(3)} \right) \\
& + |z_3|(\rho_3 + D_3) + |z_3| \left(\frac{\partial \alpha_2}{\partial x_1} \right) |(\rho_1 + D_1) + |z_3 \left(\frac{\partial \alpha_2}{\partial x_2} \right)|(\rho_3 + D_3) \\
\leq {} & \dot{V}_2 + z_3 z_4 + z_3 \left(\alpha_3 + f_3 - \frac{\partial \alpha_2}{\partial x_1}(x_2 + f_1) - \frac{\partial \alpha_2}{\partial x_2}(x_3 + f_2) - x_r^{(3)} \right) \\
& + \frac{1}{2\sigma}(z_3)^2(\rho_3 + D_3)^2 + + \frac{1}{2\sigma}(z_3)^2 \left(\frac{\partial \alpha_2}{\partial x_1} \right)^2 (\rho_1 + D_1)^2 \\
& + \frac{1}{2\sigma}(z_3)^2 \left(\frac{\partial \alpha_2}{\partial x_2} \right)^2 (\rho_3 + D_3)^2 + \frac{3\sigma}{2} \qquad (2.113)
\end{aligned}
$$

Substituting (2.109) and (2.112) into (2.113), we have

$$
\dot{V}_3 \leq -\sum_{i=1}^{3} z_i^2 + z_3 z_4 + \frac{\sigma}{2} \sum_{i=1}^{3} i. \qquad (2.114)
$$

• *Step i ($i = 4, \ldots, n$):* At this step, we choose the virtual control laws

$$
\alpha_i = -c_i z_i - z_{i-1} - f_i + \sum_{j=1}^{i-1} \frac{\partial \alpha_{i-1}}{\partial x_j}(x_{j+1} + f_j) - \frac{z_i(\rho_i + D_i)^2}{2\sigma}
$$

$$
- \sum_{j=1}^{i-1} \frac{1}{2\sigma} z_i \left(\frac{\partial \alpha_{i-1}}{\partial x_j} \right)^2 (\rho_j + D_j)^2 + x_r^{(i)}, \qquad (2.115)
$$

where c_i is a positive design constant.

At the last step n, the robust control law u is finally given by

$$
u = \alpha_n. \qquad (2.116)
$$

The Lyapunov function is defined as

$$
V_i = V_{i-1} + \frac{1}{2} z_i^2. \qquad (2.117)
$$

Then the derivative of the Lyapunov function is

$$
\dot{V}_i \leq \dot{V}_{i-1} - c_i z_i^2 - z_{i-1} z_i + z_i z_{i+1} + \frac{i\sigma}{2}. \qquad (2.118)
$$

2.3.2   Closed-loop Analysis

We now analyse the designed controller and establish the stability and tracking performance of the closed-loop system, as stated in the following theorem.

Theorem 2.3

For the closed-loop system consisting of the uncertain nonlinear system (2.99), the controller in (2.116), and the virtual controls in (2.102), (2.107), (2.112), and (2.115), all the signals in the system are bounded. Moreover, the tracking error $x(t) - x_r(t)$ converges to a bounded set and the bound of the set can be made arbitrarily small by properly adjusting the parameters $c_i(i = 1, \ldots, n)$ and σ.

Proof 2.2   Choose the final Lyapunov function

$$V = \frac{1}{2} \sum_{i=1}^{n} z_i^2 \tag{2.119}$$

From (2.116) and (2.118), one gets

$$\dot{V} \leq \dot{V}_{n-1} + z_n \left(u + f_n - \sum_{j=1}^{n-1} \frac{\partial \alpha_{n-1}}{\partial x_j}(x_{j+1} + f_j) - r^{(n)} \right)$$

$$+ |z_n|(\rho_n + D_n) + \sum_{j=1}^{n-1} \left| z_n \frac{\partial \alpha_{n-1}}{\partial x_j} \right| (\rho_j + D_j)$$

$$\leq - \sum_{i=1}^{n} c_i z_i^2 + \frac{\sigma}{2} \sum_{i=1}^{n} i$$

$$\leq - c\|z\|^2 + \Pi \tag{2.120}$$

where $c = \min\{c_1, \ldots, c_n\}$, $\Pi = \frac{n(n+1)\sigma}{4}$. It follows that $\forall \|z(t)\| \geq \sqrt{\frac{\Pi}{c}}$, $\dot{V} \leq 0$, and $z(t)$ is bounded in the set $\Lambda = \{z| \|z\| \leq \sqrt{\frac{\Pi}{c}}\}$. Therefore, the tracking error $e(t) = x(t) - x_r(t)$ will converge to the set $\{e| |e| < \sqrt{\frac{\Pi}{c}}\}$ which can be made arbitrarily small by decreasing σ or increasing c.

2.4   Robust Output Feedback Control

Now we introduce backstepping design procedures with robust output feedback for nonlinear systems described in the following form, whose nonlinearities depend only on the output y.

$$\begin{aligned}
\dot{x}_1 &= x_2 + f_1(y) + \Delta f_1(y) \\
\dot{x}_i &= x_{i+1} + f_i(y) + \Delta f_i(y) \\
\dot{x}_n &= u + f_n(y) + \Delta f_n(y) \\
y &= x_1
\end{aligned}$$
(2.121)

where $x_1, \ldots, x_n$, y and u are system states, output and input, $f_i(y) \in R^1, i = 1, \ldots, n$, are known nonlinear functions, and $\Delta f_i(y) \in R^1, i = 1, \ldots, n$, are unknown functions or uncertainties. Our control problem is to stabilize the system (2.121). For the development of control laws, the following assumption is made.

Assumption 2.8 *The unknown function $\Delta f_i(y)$ satisfies Lipschitz continuous condition, such as $|\Delta f_i(y)| \le \rho_i(y)|y|$, where $\rho_i(y) > 0$ and its $(n-i+1)$-th order derivatives are piecewise continuous.*

2.4.1 Observer Design

In order to design the desired robust output feedback control law, we rewrite the system (2.121) in the following form

$$\begin{aligned}
\dot{x} &= Ax + Bu + F(y) + \Delta F(y) \\
y &= C^T x
\end{aligned}$$
(2.122)
(2.123)

where $x = [x_1, x_2, \ldots, x_n]^T$,

$$A = \begin{bmatrix} 0 & 1 & 0 & \cdots & 0 \\ 0 & 0 & 1 & \cdots & 0 \\ \vdots & \vdots & \vdots & \ddots & \vdots \\ 0 & 0 & 0 & \cdots & 1 \\ 0 & 0 & 0 & \cdots & 0 \end{bmatrix}, \quad F(y) = \begin{bmatrix} f_1(y) \\ \vdots \\ f_n(y) \end{bmatrix}, \quad \Delta F(y) = \begin{bmatrix} \Delta f_1(y) \\ \vdots \\ \Delta f_n(y) \end{bmatrix}$$

$$B = [0, 0, \ldots, 1]^T, \quad C = [1, 0, \ldots, 0]^T.$$
(2.124)

An observer is designed to generate estimate state $\hat{x}$ as

$$\dot{\hat{x}}(t) = A\hat{x} + Bu + F(y) + K(y - \hat{y}),$$
(2.125)

where $K = [k_1, \ldots, k_n]^T$ such that $A_0 = A - KC^T$ is Hurwitz.

It can be shown that the state estimation error

$$\epsilon = x(t) - \hat{x}(t)$$
(2.126)

satisfies

$$\begin{aligned}
\dot{\epsilon} &= \dot{x}(t) - \dot{\hat{x}}(t) + \Delta F(y) \\
&= A\epsilon - K(y - \hat{y}) + \Delta F(y) \\
&= (A - KC^T)\epsilon + \Delta F(y) \\
&= A_0\epsilon + \Delta F(y)
\end{aligned}$$
(2.127)

Suppose $P \in R^{n \times n}$ is a positive definite matrix, satisfying $PA_0 + A_0^T P \le -2I$ and let

$$V_\epsilon = \epsilon^T P \epsilon \qquad (2.128)$$

It can be shown that

$$
\begin{aligned}
\dot{V}_\epsilon &= \epsilon^T (PA_0 + A_0^T P)\epsilon + 2\epsilon^T P \Delta F(y) \\
&\le -2\epsilon^T \epsilon + \epsilon^T \epsilon + \| P \Delta F(y) \|^2 \\
&\le -\epsilon^T \epsilon + \| P \rho(y) \|^2 y^2 \qquad (2.129)
\end{aligned}
$$

where $\rho(y) = [\rho_1(y), \ldots, \rho_n(y)]^T$, Assumption 2.8 and Young's inequality $2ab \le a^2 + b^2$ are used.

2.4.2 Controller Design and Stability Analysis

Note that the backstepping design starts with its output y, which is the only available system state allowed to appear in the control law. The dynamic equation of y is expressed as

$$
\begin{aligned}
\dot{y} &= x_2 + f_1(y) + \Delta f_1(y) \\
&= \hat{x}_2 + f_1(y) + \Delta f_1(y) + \epsilon_2 \qquad (2.130)
\end{aligned}
$$

Combining system (2.130) with our observer (2.125), system (2.121) is represented as

$$
\begin{aligned}
y &= \hat{x}_2 + f_1(y) + \Delta f_1(y) + \epsilon_2 & (2.131) \\
\dot{\hat{x}}_i &= \hat{x}_{i+1} + f_i(y) + k_i \epsilon_1, \ i = 2, \ldots, n-1 & (2.132) \\
\dot{\hat{x}}_n &= u + f_n(y) + k_n \epsilon_1, & (2.133)
\end{aligned}
$$

where $\epsilon_1 = y - \hat{x}_1$ is available and ϵ_2 is not available. System (2.131)–(2.133) will be our design system, whose state y and $\hat{x}_i$ are available. We now design the robust control design using the backstepping technique to regulate the output y to zero.

Theorem 2.4
For the nonlinear uncertain system (2.121) with the observer defined in (2.125), there exists a feedback robust control which guarantees global boundedness of $x(t)$ and $\hat{x}(t)$ and the asymptotic stabilization error is achieved as

$$\lim_{t \to \infty} [y(t)] = 0 \qquad (2.134)$$

One choice of the robust control is

$$u = \alpha_n \qquad (2.135)$$

where the change of coordinates z_i is defined as

$$z_1 = y \qquad (2.136)$$

$$z_i = \hat{x}_i - \alpha_{i-1}, \quad i = 2, 3, \ldots, n, \qquad (2.137)$$

and the virtual controls α_i are defined by the following recursive expressions:

$$\alpha_1 = -(c_1 + 2l_1)z_1 - f_1(y) - \sum_{j=1}^{n} \frac{1}{4l_j} \left(2 \parallel P\rho(y) \parallel^2 + (\rho_1(y))^2 \right) z_1 \qquad (2.138)$$

$$\alpha_2 = -z_1 - \left(c_2 + 2l_2 \left(\frac{\partial \alpha_1}{\partial y} \right)^2 \right) z_2 - f_2(y) - k_2\epsilon_1 + \frac{\partial \alpha_1}{\partial y}(\hat{x}_2 + f_1(y))$$

$$\qquad (2.139)$$

$$\alpha_i = -z_{i-1} - \left(c_i + 2l_i \left(\frac{\partial \alpha_{i-1}}{\partial y} \right)^2 \right) z_i - f_i(y) - k_i\epsilon_1 + \frac{\partial \alpha_{i-1}}{\partial y}(\hat{x}_2 + f_1(y))$$

$$+ \sum_{j=2}^{i-1} \frac{\partial \alpha_{i-1}}{\partial \hat{x}_j} (\hat{x}_{j+1} + f_j(y) + k_j\epsilon_1), \quad i = 3, \ldots, n \qquad (2.140)$$

$$\epsilon_1 = y - \hat{x}_1, \qquad (2.141)$$

where $c > 0$ and $l_i > 0$, $i = 1, \ldots, n$ are design parameters.

    Furthermore, the transient stabilization error performance is given by

$$\parallel z(t) \parallel_2 \ \leq \ \frac{1}{\sqrt{2c_0}} \left(\parallel z(0) \parallel^2 + \frac{1}{l_0} \parallel \epsilon(0) \parallel_P^2 \right)^{1/2} \qquad (2.142)$$

with $z(0) = [z_1(0), z_2(0), \ldots, z_n(0)]^T$, $c = \min_{1 \leq i \leq n} c_i$, $l_0 = \left(\sum_{i=1}^{n} \frac{1}{l_i} \right)^{-1}$ and $\parallel \epsilon(0) \parallel_P^2 = \epsilon(0)^T P\epsilon(0)$.

Proof: From (2.131)–(2.133) and the definitions (2.136)–(2.140), we can express the derivatives of the error variables $z_1, z_2, \ldots, z_n$ as follows:

$$
\begin{aligned}
\dot{z}_1 &= \epsilon_2 + \hat{x}_2 + f_1(y) + \Delta f_1(y) \\
&= z_2 + \alpha_1 + f_1(y) + \Delta f_1(y) + \epsilon_2 \\
&= -(c_1 + 2l_1)z_1 - \sum_{j=1}^{n} \frac{1}{4l_j} \left(2 \parallel P\rho(y) \parallel^2 + (\rho_1(y))^2 \right) z_1 \\
&\quad + \Delta f_1(y) + \epsilon_2 + z_2 \quad\quad\quad\quad\quad\quad\quad\quad (2.143)
\end{aligned}
$$

$$
\begin{aligned}
\dot{z}_2 &= \dot{\hat{x}}_2 - \dot{\alpha}_1 \\
&= z_3 + \alpha_2 + f_2(y) + k_2\epsilon_1 - \frac{\partial \alpha_1}{\partial y} \left(\hat{x}_2 + f_1(y) + \Delta f_1(y) + \epsilon_2 \right) \\
&= z_3 - z_1 - \left(c_2 + 2l_2 \left(\frac{\partial \alpha_1}{\partial y} \right)^2 \right) z_2 - \frac{\partial \alpha_1}{\partial y} (\Delta f_1(y) + \epsilon_2) \quad (2.144)
\end{aligned}
$$

$$
\dot{z}_i = z_{i+1} - z_{i-1} - \left(c_i + 2l_i \left(\frac{\partial \alpha_{i-1}}{\partial y} \right)^2 \right) z_i - \frac{\partial \alpha_{i-1}}{\partial y} (\Delta f_1(y) + \epsilon_2)
$$
$$
(2.145)
$$

The Lyapunov function V is defined as

$$
V = \sum_{i=1}^{n} \frac{1}{2} z_i^2 + \sum_{i=1}^{n} \frac{1}{2l_i} \epsilon^T P \epsilon \quad\quad\quad (2.146)
$$

Its derivative satisfies

$$
\begin{aligned}
\dot{V} &= \sum_{i=1}^{n} z_i \dot{z}_i - \sum_{i=1}^{n} \frac{1}{2l_i} \epsilon^T \epsilon + \sum_{i=1}^{n} \frac{1}{2l_i} \parallel P\Delta F(y) \parallel^2 \\
&\leq -\sum_{i=1}^{n} c_i z_i^2 - \sum_{j=1}^{2} \frac{1}{4l_j} \epsilon^T \epsilon \\
&\quad + \sum_{i=2}^{n} \left[-l_i \left(\frac{\partial \alpha_{i-1}}{\partial y} \right)^2 z_i^2 - \frac{\partial \alpha_{i-1}}{\partial y} \Delta f_1(y) z_i - \frac{1}{4l_i} (\rho_1(y) z_1)^2 \right] \\
&\quad + \sum_{i=2}^{n} \left[-l_i \left(\frac{\partial \alpha_{i-1}}{\partial y} \right)^2 z_i^2 - \frac{\partial \alpha_{i-1}}{\partial y} \epsilon_2 z_i - \frac{1}{4l_i} \epsilon^T \epsilon \right] \\
&\quad - \sum_{j=1}^{n} \frac{1}{2l_j} \parallel P\rho(y) \parallel^2 (z_1)^2 + \sum_{i=1}^{n} \frac{1}{2l_i} \parallel P\Delta F(y) \parallel^2 \\
&\leq -\sum_{i=1}^{n} c_i z_i^2 - \sum_{i=1}^{n} \frac{1}{4l_i} \epsilon^T \epsilon \quad\quad\quad\quad\quad\quad (2.147)
\end{aligned}
$$

where Young's inequality and Assumption 2.8 were used. Thus $z_1, z_2, \ldots, z_n$ and ϵ are bounded. Since $z_1 = y$ is bounded, y is also bounded. From the coordinate change (2.137), the virtual controls α_i in (2.138)–(2.140) are bounded, and following the procedure recursively, the boundedness of $x_i, \hat{x}_i, \alpha_i$ are achieved. The final control u defined in (2.135) is also bounded. Therefore, all signals in the closed-loop are globally uniformly bounded. By applying the LaSalle-Yoshizawa theorem to (2.147), it further follows that $z(t) \to 0$ as $t \to \infty$, which implies that $\lim_{t\to\infty} y(t) = 0$.

Now we derive the stabilization bound in term of L_2 norm. As shown in (2.147), the derivative of V is

$$\dot{V} \le -\sum_{i=1}^{n} c_i z_i^2 \le -c \parallel z(t) \parallel^2 \tag{2.148}$$

Since V is non-increasing, we have

$$\parallel z \parallel_2^2 = \int_0^\infty \parallel z(\tau) \parallel^2 d\tau \le \frac{1}{c_0}(V(0) - V(\infty)) \le \frac{1}{c_0}V(0) \tag{2.149}$$

where

$$V(0) = \frac{1}{2} \parallel z(0) \parallel^2 + \frac{1}{2l_0} \parallel \epsilon(0) \parallel_P^2 \tag{2.150}$$

Thus the bound in (2.142) is achieved.
△△△

2.5 Notes

This chapter gives standard procedures to design adaptive backstepping controllers and robust backstepping controllers. In the corresponding analysis parts, system stability and tracking performance are investigated. In the state-feedback control design, the results are obtained under the assumption that the full state of the system is measurable. However, for many practical problems, only a part of the state or the plant output is available for measurement. To address this issue, output-feedback control strategies using backstepping design are presented where state observers are designed to estimate the unmeasurable states.

The general design procedures and stability analysis of backstepping controllers are introduced here as preliminary knowledge for the remainder of the book. The basic design ideas and detailed analysis of the recursive backstepping techniques are not included in this chapter. Interested readers can refer to [141] and [53] for more details.

Chapter 3

Quantizers and Quantized Control Systems

When dealing with real control problems, the designer is inevitably led to face difficulties tied to limited information, such as quantized signals. Quantization technique is widely used in digital control, hybrid systems, networked systems, signal processing, simulation, embedded computing, and so on. Due to the theoretical and practical importance of the study of such dynamic systems, there has been a great deal of interest in the development of advanced control of quantized dynamic systems. In such control engineering applications, quantization is not only inevitable owing to the widespread use of digital processors that employ finite-precision arithmetic, but also useful. An important aspect is to use quantization schemes that yield sufficient precision but require a low communication rate.

This chapter introduces basic descriptions and propertties of quantizers, such as uniform quantizer, logarithmic quantizer, hysteresis-logarithmic quantizer, hystersis-uniform quantizer, and logarithmic-uniform quantizer. Four structures of quantized control systems are also presented.

3.1   Quantizers

A device or algorithmic function that performs quantization is called a quantizer, for example, analog-to-digital converter. A quantizer can be mathematically modeled as

DOI: 10.1201/9781003176626-3

a discontinuous map from a continuous region to a discrete set of numbers. In this section, the analytical expressions of quantizers are presented.

3.1.1 Uniform Quantizer

A uniform quantizer has equal quantization levels and is optimal for uniformly distributed signals. It is the most commonly used quantizer in signal processing. The analytical expression of uniform quantizer can be described as

$$q_u(\chi) = \begin{cases} \chi_i sgn(\chi), & \chi_i - \frac{l}{2} < |\chi| \le \chi_i + \frac{l}{2} \\ 0, & |\chi| \le \chi_0 \end{cases}, \tag{3.1}$$

where $\chi_0 > 0$ and $\chi_1 = \chi_0 + \frac{l}{2}$, $\chi_i = \chi_{i-1} + l$ with $i = 2, \ldots$, and l is the length of the quantization interval. $q_u(\chi)$ is in the set $U = \{0, \pm \chi_i\}$. Clearly the uniform quantization error is bounded by a positive constant $\Delta = \max\{\chi_0, l\}$. The map of the uniform quantizer $q_u(\chi)$ for $\chi > 0$ is shown in Figure 3.1.

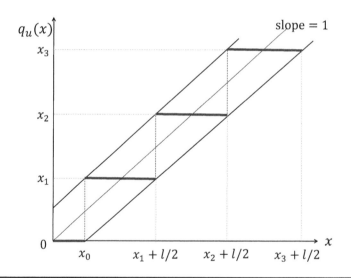

Figure 3.1 The map of uniform quantizer $q_u(\chi)$

Some other forms of uniform quantizers and their applications in distributed averaging, vehicle control and robotic actuator control can be found in [11, 47, 49, 76]. However, many signals have non-uniform distribution, such as audio signals. For such signals, a uniform quantizer may be wasteful and a non-uniform quantizer is a better choice, such as logarithmic quantizer.

3.1.2 Logarithmic Quantizer

The logarithmic quantizer belongs to non-uniform quantization in which the quantization levels are unequal. The analytical expression of logarithmic quantizer is given below as in [80, 145].

$$q_{log}(\chi) = \begin{cases} \chi_i sgn(\chi), & \frac{\chi_i}{1+\delta} < |\chi| \leq \frac{\chi_i}{1-\delta} \\ 0, & |\chi| \leq \frac{\chi_0}{1+\delta} \end{cases}, \tag{3.2}$$

where $\chi_i = \rho^{(1-i)}$, $\chi_0 > 0$, $0 < \delta < 1$, and $\rho = \frac{1-\delta}{1+\delta}$ are quantization parameters. $q_{log}(\chi)$ is in the set $U = \{0, \pm \chi_i\}$. The map of the logarithmic quantizer (3.2) for $\chi > 0$ is shown in Figure 3.2.

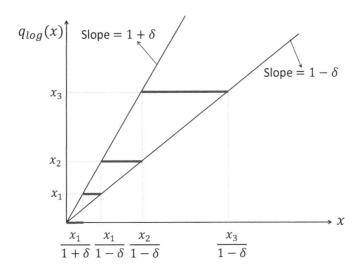

Figure 3.2 The map of logarithmic quantizer $q_{log}(\chi)$

A logarithmic quantizer is useful where the signals are more concentrated near the equilibrium or have higher resolution around the equilibrium. There are several advantages of using logarithmic quantization, such as the coarsest quantization density and low communication capacity. Logarithmic quantization has been used in many applications such as digitizing analog signals, color quantization in image processing, and son on.

Remark 3.1   The parameter ρ is considered as a measure of quantization density. The smaller the ρ is, the coarser the quantizer is. When ρ approaches to zero, δ approaches to 1, then $q_{log}(\chi)$ will have fewer quantization levels as χ ranges over that interval. Such kind of quantizers is the coarsest quantizer which minimizes the average rate of communication instances and is easy to implement.

3.1.3 Hysteresis-Uniform Quantizer

One of the difficulties in control of continuous-time quantized systems is that there may be chattering or undesirable switching using the quantized signals. This is not desirable from a networked control viewpoint since transmission of such signals requires infinite bandwidth. Moreover, mathematically, chattering brings sensitive issues regarding the existence of a solution to the system. To prevent chattering and switching phenomena, a hysteresis-type mechanism can be introduced in the quantization, such as hysteresis-uniform quantizer and hysteresis-logarithmic quantizer.

The analytical expression of hysteresis-uniform quantizer is

$$
q_{hu}(\chi) = \begin{cases}
\chi_i sgn(\chi), & \chi_i - \frac{l}{2} - h < |\chi| \le x_i - \frac{l}{2} + h \\
& \text{and } \dot{\chi} < 0, \text{or} \\
& \chi_i + \frac{l}{2} - h < |\chi| < \chi_i + \frac{l}{2} + h \\
& \text{and } \dot{\chi} > 0, \text{or} \\
& \chi_i - \frac{l}{2} + h \le |\chi| \le \chi_i + \frac{l}{2} - h \\
0, & -\chi_0 - h \le \chi \le \chi_0 + h \\
q_{hu}(\chi(t^-)), & \dot{\chi} = 0
\end{cases} \tag{3.3}
$$

where $\chi_0 = \frac{l}{2}$ and $\chi_{i+1} = \chi_i + l$, l is the length of the quantization interval, $h = p_h l$ is the hysteresis width constant and $0 < p_h \le 0.5$ is hysteresis percentage, $q_{hu}(\chi)$ is in the set $U = \{0, \pm \chi_i\}$, χ_0 determines the size of the dead-zone for $q_{hu}(\chi)$. The map of the hysteresis-uniform quantizer $q_{hu}(\chi)$ for $\chi > 0$ is shown in Figure 3.3.

Remark 3.2   Compared with the uniform quantizer in (3.1), the hysteresis-uniform quantizer in (3.3) has additional quantization levels. Two quantizers of the same coarseness but with different quantization levels are employed and are switched when the output value of the quantizer changes. Whenever the quantized signals make a transition from one value to another, some dwell time will elapse before a new transition can occur as shown in Figure 3.3, which will avoid chattering. This can be seen as a way to add hysteresis to uniform quantization.

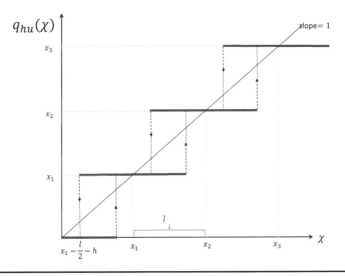

Figure 3.3 The map of hysteresis-uniform quantizer $q_{hu}(\chi)$

3.1.4 Hysteresis-Logarithmic Quantizer

The analytical expression of hysteresis-logarithmic quantizer is given below as in [20, 147]

$$
q_{hys}(\chi) =
\begin{cases}
\chi_i \mathrm{sgn}(\chi), & \frac{\chi_i}{1+\delta} < |\chi| \le \chi_i, \dot{\chi} < 0, \text{or} \\
& \chi_i < |\chi| \le \frac{\chi_i}{1-\delta}, \dot{\chi} > 0 \\
\chi_i(1+\delta)\mathrm{sgn}(\chi), & \chi_i < |\chi| \le \frac{\chi_i}{1-\delta}, \dot{\chi} < 0, \text{or} \\
& \frac{\chi_i}{1-\delta} < |\chi| \le \frac{\chi_i(1+\delta)}{(1-\delta)}, \dot{\chi} > 0 \\
0, & |\chi| < \frac{\chi_0}{1+\delta}, \dot{\chi} < 0, \text{or} \\
& \frac{\chi_0}{1+\delta} \le \chi \le \chi_0, \dot{\chi} > 0, \\
q_{hys}(\chi(t^-)), & \dot{\chi} = 0
\end{cases}
\tag{3.4}
$$

where $\chi_i = \rho^{(1-i)}$, $\chi_0 > 0$, $0 < \delta < 1$, and $\rho = \frac{1-\delta}{1+\delta}$ are quantization parameters. $q_{hys}(\chi)$ is in the set $U = \{0, \pm\chi_i, \pm\chi_i(1+\delta)\}$. The map of hysteresis-logrithmic quantizer (3.4) is shown in Fig. 3.4.

Remark 3.3   Compared with the logarithmic quantizer in (3.2), the hysteresis-logarithmic quantizer in (3.4) has additional quantization levels, which are used to avoid chattering. Whenever $q_{hys}(\chi)$ makes a transition from one value to another, some dwell time will elapse before a new transition can occur, as shown in Figure 3.4. This can be seen as a way to add hysteresis to the quantized system. The effectiveness of hysteresis-logarithmic quantizers have been verified in many practical systems, such as spacecraft systems [94, 121], quadrotor systems [133], and robotic manipulator systems [54]. More information about hysteresis-logarithmic quantizers can be found in [12, 146, 147].

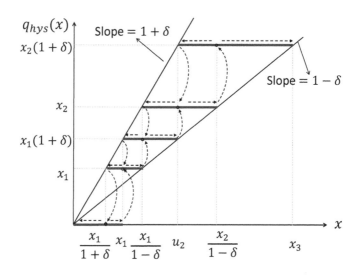

Figure 3.4 The map of hys-log quantizer $q_{hys}(\chi)$

3.1.5 Logarithmic-Uniform Quantizer

As pointed out in the logarithmic quantizer, the quantization level becomes coarser as the magnitude of the signal gets bigger (away from the origin), which results in unnecessary large quantization error. To overcome this problem, a logarithmic-uniform quantizer combining a uniform quantizer and a logarithmic quantizer, which was developed in [127], is modeled as

$$
q_{lu}(\chi(t)) = \begin{cases} q_{log}(\chi_{th}) + q_u\left(\chi - \chi_{th}\right), & |\chi| \geq \chi_{th} \\ q_{log}(\chi) & |\chi| < \chi_{th} \end{cases} \tag{3.5}
$$

where χ_{th} is a positive constant specified by designer denoting the threshold to switch between the logarithmic and uniform quantizer. q_u and q_{log} represent a uniform quantizer defined in (3.1) and a logarithmic quantizer defined in (3.2). The map of the quantizer $q_{lu}(\chi)$ is shown in Figure 3.5.

Remark 3.4   With the logarithmic-uniform quantizer $q_{lu}(\chi)$, it is guaranteed that the quantization error $|q_{lu}(\chi) - \chi|$ for $|\chi| > \chi_{th}$ remains the same as that of the logarithmic quantizer $q_{log}(\chi)$ when $|\chi| = \chi_{th}$. This can be considered as that a saturation level is introduced to the quantization error of the traditional logarithmic quantizer. Note that χ_{th} is a user-defined parameter denoting the trade-off between system performances and communication burden, and it can be chosen according to the practical situations.

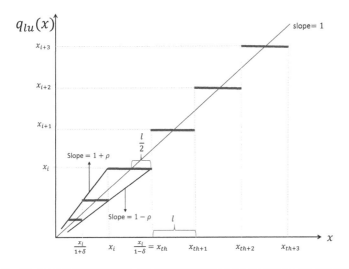

Figure 3.5 The map of logarithmic-uniform quantizer $q_{lu}(\chi)$

3.2 Properties

The difference between an input value and its quantized value is referred to as quantization error. Based on the boundedness of the quantization error, there are two types of quantization: bounded property and sector-bounded property.

3.2.1 Bounded Property

A bounded quantizer has a bounded quantization error which meets the following property:

$$|q(\chi) - \chi| \leq \Delta \qquad (3.6)$$

where $\Delta > 0$ is the bound of quantization error. It can be shown that the uniform quantizer in (3.1), and the hysteresis-uniform quantizer in (3.3) and the logarithmic-uniform quantizer in (3.5) illustrated above have the property (3.6).

- For a uniform quantizer in (3.1), the quantization error is bounded by a positive constant $\Delta = \max\{\chi_0, \, l\}$.

- For a hysteresis-uniform quantizer in (3.3), the quantization error is bounded by a positve constant $\Delta \geq \frac{1}{2}l + h$.

- For a logarithmic-uniform quantizer in (3.5), the quantization error is bounded by a positive constant $\Delta \geq \frac{1}{2}l$.

3.2.2 Sector-Bounded Property

A sector-bounded quantizer has the following sector-bounded property:

$$|q(\chi) - \chi| \leq \delta \|\chi\| + \Delta, \tag{3.7}$$

where $0 \leq \delta < 1$ and $\Delta > 0$ are quantization parameters. It can be shown that logarithmic quantizer in (3.2) and hysteresis-logarithmic quantizer in (3.4) satisfy the sector-bounded property in (3.7).

■ In logarithmic quantizer in (3.2), $|q_{log}(\chi) - \chi| \leq \delta \|\chi\| + \Delta$, where $0 < \delta < 1$ is the quantization parameter and $\Delta = \frac{\chi_0}{1+\delta} > 0$ determines the size of the dead-zone for $q_{log}(\chi)$.

■ In hysteresis-logrithmic quantizer (3.4), $|q_{hys}(\chi) - \chi| \leq \delta \|\chi\| + \Delta$, where $0 < \delta < 1$ is the quantization parameter and $\Delta = \chi_0$ determines the size of the dead-zone for $q_{hys}(\chi)$.

It is noted that the quantization error of bounded quantizer is bounded by a constant. By contrast, the quantization error of sector-bounded quantizer depends on the input of the quantizer, which cannot be ensured bounded automatically. This constitutes the main challenge to handle the effects of sector-bounded quantizer in the stability analysis.

3.3 Quantized Control Systems

In a networked control system, the sensors of a plant transmit their measured signals to the controllers through the sensor-to-controller (S-C) channel, while the controllers send the computed control information to the actuators through the controller-to-actuator (C-A) channel. As these communication channels may be shared by different systems nodes and their communication bandwidth is limited (especially for wireless communication networks), it is desired to reduce redundant signal transmissions over the S-C and C-A channels without affecting control performance.

Signal quantization, realized through Quantizers, is one of the most important techniques in decreasing signal transmission burden, and it can be applied to the S-C channels and/or the C-A channels. The integration of signal quantizations transforms traditional networked systems to quantized control systems. According to different system configurations in which signal quantization is used, typical quantized control systems can be divided into four categories:

■ System with input quantization. This category aims to reduce the communication burden in C-A channels [50, 100, 132].

■ System with state (or output) quantization. It aims to reduce the communication burden in S-C channels [28, 29, 44, 67].

- System with both input and state quantization. Communication burden can be reduced in both S-C and C-A channels [24].

- System with both input and output quantization. Communication burden can be reduced in both S-C and C-A channels [17, 23, 130].

Figures 3.6–3.9 show the block diagrams of above four categories of quantized control systems. Figures 3.6 shows the block diagram of feedback control systems with input quantization, where the control signal u is quantized (via input quantizer) and the quantized control signal u^q is sent to the system.

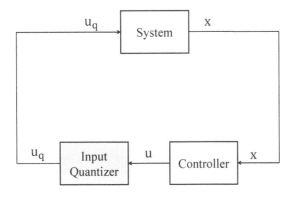

Figure 3.6 System with input quantization

Figures 3.7 shows the block diagram of feedback control systems with state quantization, where all state signals are quantized (via state quantizer) and sent to the controller. For the system with quantized states, the state x is not available and only the quantized state x_q can be used in the controller u.

Figures 3.8 shows the block diagram of feedback control systems with input quantization and state quantization, where all states are quantized (via state quantizer) and sent to the controller, and similarly the control signal is quantized (via input quantizer) and sent to the system.

Figures 3.9 shows the block diagram of feedback control systems with input quantization and output quantization, where only output signals y are quantized (via output quantizer) and sent to the controller, and similarly the control signal is quantized (via input quantizer) and sent to the system. In such systems, only the quantized output signals y_q can be used in the designed controller u.

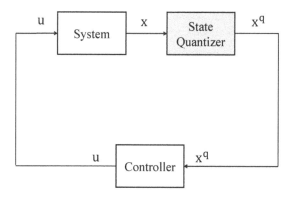

Figure 3.7 System with state quantization

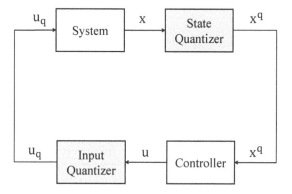

Figure 3.8 System with input and state quantization

3.4 Notes

In this chapter, the basic descriptions and properties of quantizers are introduced as preliminary knowledge for the remainder of this book. Four types of quantized control systems are generally presented: a system with input quantization, a system with state quantization, a system with input and state quantization, and a system with input and output quantization. In the remainder of this book, we will present series of innovative technologies and research results on adaptive control of these four types of quantized control systems with different quantizers, including adaptive backstepping controller design, stability analysis, compensation for effects of quantization, and improvement of system tracking and transient performance.

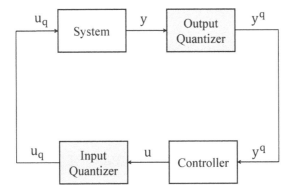

Figure 3.9 System with both input and output quantization

INPUT QUANTIZATION COMPENSATION

Chapter 4

Adaptive Stabilization of Nonlinear Uncertain Systems with Input Quantization

In this chapter, we study a stabilization problem for a class of strict-feedback nonlinear systems, where the input signal takes quantized values from a hysteresis-logarithmic quantizer. The considered nonlinear systems satisfy the Lipschitz condition. The control design is achieved by using the backstepping technique and a guideline is derived to select the parameters of the quantizer. The designed controller together with the quantizer ensures the stability of the closed-loop system in the sense of signal boundedness.

4.1 Introduction

As discussed in Chapter 3, the quantization technique is widely used in digital control, hybrid systems, and networked systems due to its theoretical and practical importance. Much attention has been paid to quantized feedback control, in order to understand the required quantization density or information rate in stability analysis. Research on stabilization of linear and nonlinear systems with quantized control signals has received great attention, see for examples, [23, 45, 61, 62, 75, 81, 98]. The systems considered in the above references are completely known.

In practice, it is often required to consider the case where the plant to be controlled is uncertain. Quantized control of systems with uncertainties has been studied by using robust approaches, see for examples, [20,66,67,67,80,129]. As well known, adaptive control is a useful and important approach to deal with system uncertainties due to its ability to provide online estimations of unknown system parameters with measurements. However, results based on the adaptive control approach are still very limited. It is noted that adaptive control schemes with quantized input have been reported in [33,34,95]. In [33,95], adaptive control for linear uncertain systems with input quantization was studied. In [34], adaptive quantized control of nonlinear systems was considered, where the idea of constructing a hysteretic type of input quantization was originally introduced. However, the stability condition in [33,34] depends on the control signal, which is hard to be checked in advance as the control signal is only available after the controller is put in operation. This limitation is removed in [142, 147], where an adaptive backstepping control scheme is proposed for uncertain strict-feedback nonlinear systems.

In this chapter, an adaptive backstepping feedback the control scheme is developed for a class of strict-feedback nonlinear systems preceded by quantized input signal by considering a stabilization problem. A hysteresis-logarithmic quantizer is studied. The quantization parameters will be chosen based on a derived inequality related to the given controller design parameters and certain system parameters. In this way, stability in the sense of ultimate boundedness is achieved by choosing suitable quantization parameters and design parameters, which can be easily verified in advance. Thus the proposed scheme relaxes the stability condition in [33, 34]. It also ensures that the ultimate stabilization error is proportional to a design parameter and thus adjustable. Simulation results illustrate the effectiveness of our proposed scheme.

4.2   System Model

Consider a control system over the network with control input quantization. The control input is quantized and then coded in the coder to be sent over the network. We assume that the network is noiseless, so the quantized input signal is recovered in the decoder and applied to the plant. In this chapter, a class of nonlinear plants is considered in the following parametric strict-feedback form as in [53,71].

$$
\begin{aligned}
\dot{x}_1 &= x_2 + \psi_1(x_1) \\
\dot{x}_2 &= x_3 + \psi_2(x_1, x_2) \\
&\ \ \vdots \qquad \vdots \\
\dot{x}_{n-1} &= x_n + \psi_{n-1}(x_1, \ldots, x_{n-1}) \\
\dot{x}_n &= q(u(t)) + \phi^T(x)\theta + \psi_n(x)
\end{aligned} \tag{4.1}
$$

where $x(t) = [x_1(t), \ldots, x_n(t)]^T \in \Re^n$ and $q(u(t)) \in \Re^1$ are the states and input of the system, respectively, $\bar{x}_i(t) = [x_1(t), \ldots, x_i(t)]^T \in \Re^i$, the vector $\theta \in \Re^r$

is constant and unknown, $\phi \in \Re^r, \psi_i \in \Re^1$, $i = 1, \ldots, n$ are known nonlinear functions and differentiable. The input $q(u(t))$ represents the quantizer and takes the quantized values, where $u(t) \in \Re^1$ is the control input signal to be quantized at the encoder side. For this class of nonlinear systems, we assume that the existence and uniqueness of solution are satisfied.

For the development of control laws, the following assumptions are also made.

Assumption 4.1 *The nonlinear functions ϕ and ψ_i satisfy the global Lipschitz continuity condition such that*

$$\| \phi(t,x) - \phi(t,y) \| \leq L_\phi \| x - y \| \tag{4.2}$$

$$\| \frac{\partial \phi}{\partial x}(t,x) \| \leq L_\phi$$

$$\| \psi_i(t,\bar{x}_i) - \psi_i(t,\bar{y}_i) \| \leq L_{\psi_i} \| \bar{x}_i - \bar{y}_i \| \tag{4.3}$$

$$\| \frac{\partial \psi_i}{\partial \bar{x}_i}(t,\bar{x}_i) \| \leq L_{\psi_i}$$

$\forall x,y \in \Re^n$, $\forall \bar{x}_i, \bar{y}_i \in \Re^i$, *where L_ϕ and L_{ψ_i} are known constants.*

Assumption 4.2 *The unknown parameter vector θ is within a known compact convex set C such that $\| \theta_a - \theta_b \| \leq \theta_M$ for any $\theta_a, \theta_b \in C$ and a constant θ_M.*

The control objective is to design a backstepping feedback control law for $u(t)$ to ensure that all closed-loop signals are bounded and the ultimate stabilization error is within an adjustable bound.

4.3 Quantizer

In this chapter, a hysteresis-logarithmic quantizer $q(u(t))$ is used to avoid chattering and its mathematical description is given in (3.4) in Chapter 3. It is noted that $q(u(t))$ is in the set $U = \{0, \pm u_i, \pm u_i(1+\delta)\}$, where $u_i = \rho^{(1-i)} u_{min}$ with integer $i = 1, 2, \ldots$ and parameters $u_{min} > 0$ and $0 < \rho < 1$, $\delta = \frac{1-\rho}{1+\rho}$. The map of hysteresis-logarithmic quantizer $q(u(t))$ for $u > 0$ is shown in Figure 4.1. The discussions of this kind of quantizer is made in Chapter 3.

In order to propose a suitable control scheme, we decompose the hysteresis-logarithmic quantizer $q(u(t))$ into a linear part and a nonlinear part as follows.

$$q(u(t)) = u(t) + d(t) \tag{4.4}$$

where $d(t) = q(u(t)) - u(t) \in \Re^1$. Regarding the nonlinearity $d(t)$, we have the following lemma.

Lemma 4.1
The nonlinearity $d(t)$ satisfies the following inequality.

$$d^2(t) \leq \delta^2 u^2, \ \forall \, |u| \geq u_{min}. \tag{4.5}$$

$$d^2(t) \leq u_{min}^2, \ \forall \, |u| \leq u_{min}. \tag{4.6}$$

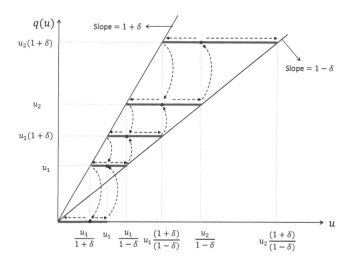

Figure 4.1 The map of $q(u)$ for $u > 0$

Proof 4.1   From Figure 4.1, we can get that for $u \geq u_{min}$

$$(1 - \delta)u \;\leq\; q(u) \leq (1 + \delta)u \tag{4.7}$$
$$|q(u) - u| \;\leq\; \delta u \tag{4.8}$$

Similarly for $u \leq -u_{min}$, it can be shown that $|q(u) - u| \leq -\delta u$. Thus, we have $|q(u) - u| \leq \delta|u|$ for $|u| \geq u_{min}$, which implies the property (4.5). For $|u| \leq u_{min}$, $q(u) = 0$ from the definition. So the property (4.6) is derived directly.

4.4   Design of Adaptive Controller

In this section, we will design adaptive backstepping feedback control laws for the nonlinear uncertain system (4.1) where the input is quantized by a hysteresis-logarithmic quantizer. The objective is to design an adaptive backstepping controller and a quantization rule (i.e. a guideline to choose the quantization parameters) for $u(t)$ to reduce the information to be sent over the communication channel in the presence of system uncertainties. We have no apriori information on how fine the quantizer should be to give a stable closed-loop system. To achieve the objective, a guideline of choosing the quantization parameters is derived. For nonlinear uncertain system (4.1), the number of design steps required is equal to n. At each step, an error variable z_i and a stabilizing function α_i is generated. Finally, the control u and a parameter estimate $\hat{\theta}$ are developed.

Introduce the change of coordinates

$$z_1 = x_1 \tag{4.9}$$

$$z_i = x_i - \alpha_{i-1}, \ i = 2, 3, \ldots, n \tag{4.10}$$

where α_i are virtual controllers. The design procedure is elaborated in the following steps.

Step i, ($i = 1, \ldots, n - 1$). The design for the first $n - 1$ susbsytems follows the backstepping design procedure in Chapter 2. Design the stabilizing function α_i as

$$\alpha_i = -(c_i + 1)z_i - \psi_i + \sum_{j=1}^{i-1} \frac{\partial \alpha_{i-1}}{\partial x_j}(x_{j+1} + \psi_j) \tag{4.11}$$

where c_i is a positive constant.

Step n. In the last step n, the actual control input u appears and is at our disposal. The z_n dynamics is given as

$$\dot{z}_n = q(u) + \psi_n - \sum_{j=1}^{n-1} \frac{\partial \alpha_{i-1}}{\partial x_j}(x_{j+1} + \psi_j) + \theta^T \phi$$

$$= u(t) + d(t) + \psi_n - \sum_{j=1}^{n-1} \frac{\partial \alpha_{i-1}}{\partial x_j}(x_{j+1} + \psi_j) + \theta^T \phi \tag{4.12}$$

We choose the following Lyapunov function

$$U_n = \sum_{i=1}^{n} \frac{1}{2}z_i^2 + \frac{1}{2}\tilde{\theta}^T \Gamma^{-1}\tilde{\theta} \tag{4.13}$$

where Γ is a positive definite matrix, and $\tilde{\theta} = \theta - \hat{\theta}$. Then its derivative is given by

$$\dot{U}_n = \sum_{i=1}^{n} \frac{1}{2}z_i\dot{z}_i + \tilde{\theta}^T \Gamma^{-1}\dot{\tilde{\theta}}$$

$$\leq -\sum_{i=1}^{n-1} c_i z_i^2 + \frac{1}{2}z_n^2 + \tilde{\theta}^T \Gamma^{-1}\dot{\tilde{\theta}}$$

$$+ z_n\left(u(t) + d(t) + \psi_n - \sum_{j=1}^{n-1} \frac{\partial \alpha_{n-1}}{\partial x_j}(x_{j+1} + \psi_j) + \theta^T \phi\right)$$

$$\leq -\sum_{i=1}^{n-1} c_i z_i^2 + z_n d(t) + \tilde{\theta}^T \Gamma^{-1}\left(\Gamma\phi z_n - \dot{\hat{\theta}}\right)$$

$$+ z_n\left(u(t) + \frac{1}{2}z_n + \psi_n - \sum_{j=1}^{n-1} \frac{\partial \alpha_{n-1}}{\partial x_j}(x_{j+1} + \psi_j) + \hat{\theta}^T \phi\right) \tag{4.14}$$

We are finally at the position to design a control law for u and update law for $\dot{\hat{\theta}}$ as

$$u = -(c_n + \frac{1}{2} + k_d)z_n - \psi_n + \sum_{j=1}^{n-1} \frac{\partial \alpha_{n-1}}{\partial x_j}(x_{j+1} + \psi_j) - \hat{\theta}^T \phi \quad (4.15)$$

$$\dot{\hat{\theta}} = Proj\{\Gamma \phi z_n\} \quad (4.16)$$

where c_n and k_d are positive constants. $Proj(.)$ is the projection operator given in Appendix C, which ensures that $\|\hat{\theta}\| \leq \theta_M$ for a given constant θ_M. Using the property of the projection operator $-\tilde{\theta}^T \Gamma^{-1} Proj(\tau) \leq -\tilde{\theta}^T \Gamma^{-1} \tau$ in Appendix C and from (4.14)–(4.16), the derivative of U_n satisfies

$$\dot{U}_n \leq -\sum_{i=1}^{n} c_i z_i^2 + z_n d(t) - k_d z_n^2 + \tilde{\theta}^T \Gamma^{-1} \left(\Gamma \phi z_n - Proj\{\Gamma \phi z_n\} \right)$$

$$\leq -\sum_{i=1}^{n} c_i z_i^2 + z_n d(t) - k_d z_n^2 \quad (4.17)$$

4.5 Stability Analysis

Now we will find an appropriate quantization parameter δ to ensure the boundedness of U_n. To do this, we need the following lemma.

Lemma 4.2
The virtual control law α_i, $i = 1, \ldots, n-1$ and the final control $u(t)$ satisfy the Lipschitz condition, namely

$$|\alpha_i| \leq k_{\alpha_i} \|\bar{z}_i(t)\| \quad (4.18)$$

$$|u(t)| \leq k_u \|z(t)\| \quad (4.19)$$

for positive constants k_{α_i} and k_u which depend on the given design parameters and certain system parameters, $\bar{z}_i(t) = [z_1, z_2, \ldots, z_i]^T$, $z(t) = [z_1, z_2, \ldots, z_n]^T$.

Proof 4.2 From Assumption 4.1, the functions ϕ and ψ_i satisfy the Lipschitz continuity condition. Since $\alpha_1 = -(c_1 + \frac{1}{2})z_1 - \psi_1$, it can be shown that

$$|\alpha_1| \leq (c_1 + \frac{1}{2})|z_1| + L_{\psi_1}|z_1| = k_{\alpha_1}|z_1| \quad (4.20)$$

where $k_{\alpha_1} = c_1 + \frac{1}{2} + L_{\psi_1}$.

$$|\frac{\partial \alpha_1}{\partial x_1}| \leq (c_1 + \frac{1}{2}) + |\frac{\partial \psi_1}{\partial x_1}| \leq (c_1 + \frac{1}{2}) + L_{\psi_1} = k_{\alpha_1} \quad (4.21)$$

$$
\begin{aligned}
|x_2| &\leq |z_2 + \alpha_1| \\
&\leq |z_2| + k_{\alpha_1}|z_1| \leq (2k_{\alpha_1}^2 z_1^2 + 2z_2^2)^{1/2} \\
&\leq \max(\sqrt{2k_{\alpha_1}}, \sqrt{2}) \, \| \bar{z}_2 \| \\
&= k_{x2} \, \| \bar{z}_2 \|,
\end{aligned}
\quad (4.22)
$$

where $k_{x_2} = \max(\sqrt{2k_{\alpha_1}}, \sqrt{2})$. Following the similar procedure based on $\alpha_i = -(c_i + 1)z_i - \psi_i + \sum_{j=1}^{i-1} \frac{\partial \alpha_{i-1}}{\partial x_j}(x_{j+1} + \psi_j)$ in (4.11), we have

$$
\begin{aligned}
|\alpha_i| &\leq (c_i + 1)|z_i| + |\sum_{j=1}^{i-1} \frac{\partial \alpha_{i-1}}{\partial x_j}(x_{j+1} + \psi_j)| + |\psi_i| \\
&\leq (c_i + 1)|z_i| + L_{\psi_i} \, \| \bar{x}_i \| + |\sum_{j=1}^{i-1} k_{\alpha_{i-1}}(|x_{j+1}| + L_{\psi_j} \, \| \bar{x}_j \|)| \\
&\leq k_{\alpha_i} \, \| \bar{z}_i \|
\end{aligned}
\quad (4.23)
$$

$$|\frac{\partial \alpha_i}{\partial x_i}| \leq k_{\alpha_i} \quad (4.24)$$

$$|x_i| \leq |z_i + \alpha_{i-1}| \leq |z_i| + k_{\alpha_i} \, \| \bar{z}_i \| \leq k_{x_i} \, \| \bar{z}_i \| \quad (4.25)$$

$$
\begin{aligned}
\| \bar{x}_i \| &= (\sum_{j=1}^{i} x_j^2)^{1/2} \\
&\leq (\sum_{j=1}^{i} k_{xj} \, \| \bar{z}_j \|^2)^{1/2} \leq (\sum_{j=1}^{i} k_{xj})^{1/2} \, \| \bar{z}_i \| \leq k_{\bar{x}_i} \, \| \bar{z}_i \|
\end{aligned}
\quad (4.26)
$$

From $\| \hat{\theta} \| \leq \theta_M$ and according to Assumption 1 and (4.23)–(4.26), the final control u in (4.15) satisfies

$$|u(t)| \leq k_u \, \| z(t) \|. \quad (4.27)$$

We are now at the position to state our main results in the following theorem.

Theorem 4.1
Consider the closed-loop adaptive system consisting of plant (4.1), the adaptive backstepping controller (4.15) with virtual control laws (4.11), parameter estimator with updating law (4.16), and the hysteresis-logarithmic quantizer. The boundedness of all the signals in the system is ensured if the quantized parameter δ satisfies

$$\beta - \delta k_u \geq \epsilon > 0. \quad (4.28)$$

where $\beta = min\{c_1, c_2, \ldots, c_n\}$ and ϵ is a positive conatant. Furthermore, the stabilization error is ultimately bounded as follows:

$$\| z(t) \| \leq B, \text{ where } B = \sqrt{\frac{u_{min}^2}{4k_d\epsilon}}. \tag{4.29}$$

Proof 4.3 We consider two cases to get a bound for $\dot{U}_n$.

Case 1. $|u(t)| \leq u_{min}$. Using $|d(t)| \leq u_{min}$ in (4.6), we have from (4.17) that

$$
\begin{aligned}
\dot{U}_n & \leq -\sum_{i=1}^{n} c_i z_i^2 + |z_n| u_{min} - k_d z_n^2 \\
& \leq -\sum_{i=1}^{n} c_i z_i^2 + \frac{1}{4k_d} u_{min}^2 \\
& \leq -\beta \| z(t) \|^2 + \frac{1}{4k_d} u_{min}^2, \ \forall \, |u(t)| \leq u_{min}. \tag{4.30}
\end{aligned}
$$

Case 2. $|u(t)| > u_{min}$. Using the property that $d^2(t) \leq \delta^2 u^2$ in (4.5), (4.17) can be written as

$$\dot{U}_n \leq -\sum_{i=1}^{n} c_i z_i^2 + \delta |u(t)| \| z \| \tag{4.31}$$

From (4.18)–(4.19) in Lemma 4.2, we have

$$
\begin{aligned}
\sum_{i=1}^{n} c_i z_i^2 - \delta |u| \| z \| & \geq \beta \| z(t) \|^2 - \delta k_u \| z(t) \|^2 \\
& = (\beta - \delta k_u) \| z(t) \|^2, \ \forall \, |u(t)| > u_{min} \tag{4.32}
\end{aligned}
$$

If the quantization parameter δ is chosen to satisfy (4.28), the derivative of U_n in (4.31) satisfies

$$
\begin{aligned}
\dot{U}_n & \leq -\sum_{i=1}^{n} c_i z_i^2 + \delta |u(t)| \| z \| \\
& \leq -(\beta - \delta k_u) \| z(t) \|^2 \\
& \leq -\epsilon \| z(t) \|^2 < 0, \ \forall \, |u(t)| > u_{min}. \tag{4.33}
\end{aligned}
$$

Combining the two cases from (4.30) and (4.33), we obtain, for all $t \geq 0$, that

$$\dot{U}_n \leq -\epsilon \| z(t) \|^2 + \frac{1}{4k_d} u_{min}^2. \tag{4.34}$$

It follows that $\dot{U}_n < 0$, $\forall \, \| z(t) \| > \sqrt{\frac{u_{min}^2}{4k_d\epsilon}}$. Then the ultimate bound of $z(t)$ satisfies (4.29). From (4.34) as well as the projection operation (4.16), $z_1, z_2, \ldots, z_n$

and $\hat{\theta}$ are bounded under condition (4.28). This further implies that x_1 is bounded. The boundedness of x_2 follows from the boundedness of α_1 and the fact that $x_2 = z_2 + \alpha_1$. Similarly, the boundedness of x_i $(i = 3, \ldots, n)$ can be ensured from the boundedness of α_i in (4.11) and the fact that $x_i = z_i + \alpha_{i-1}$. Combining this with (4.15) or (4.19), we conclude that the control $u(t)$ is also bounded. Thus all the signals are bounded.

Remark 4.1   The ultimate stabilization error is proportional to u_{min}. As we can choose u_{min}, so such a stabilization error is adjustable and can be made arbitrarily small.

Remark 4.2   The controller designed in this section achieves the goal of stabilization with quantized input signal. The proof of these properties is a direct consequence of the recursive procedure, because a Lyapunov function is constructed for the entire system including the parameter estimates.

Remark 4.3   Note that the choice of δ is arbitrary so long as (4.28) holds for the given design parameters c_i. So, (4.28) can be considered as a guideline to choose this quantization parameter. Based on Theorem 4.1, the required number of quantization levels is finite since the control signal (4.15) is bounded and δ is bounded.

Remark 4.4   As stated in Theorem 4.1, the quantized parameter δ is chosen to satisfy (4.28). To do this, k_u should be known. It is noted that k_u depends on L_ϕ, L_{ψ_i} and other system parameters in Lemma 4.2 to guarantee that $u(t)$ is bounded by $k_u \parallel z(t) \parallel$. It follows that L_ϕ and L_{ψ_i} should be known which are assumed in Assumption 4.1.

4.6   An Illustrative Example

In this section we consider a nonlinear system with a quantized input as follows.

$$\ddot{x} + \theta \sin(\dot{x}) + \tanh(x) = q(u) \tag{4.35}$$

where $q(u)$ represents a hysteresis-logarithmic quantizer as in (3.4), parameters θ is unknown, $u_{min} = 0.02$. The objective is to design a quantized control input for u to stabilize the system. The actual parameter value is chosen as $\theta = 1$ for simulations. In the simulations, we choose $c_1 = c_2 = 1, \gamma = 1$. The quantization parameter is chosen as $\delta = 0.2$ which satisfies the stabilizing condition (4.28), where $k_u = 5$. The initial states and parameter are chosen as $x(0) = \dot{x}(0) = 0.5$, and $\hat{\theta}(0) = 0.82$. Figures 4.2–4.4 show the trajectories of states x and $\dot{x}$, the estimated parameter $\hat{\theta}$, the actual input $u(t)$, and the quantized input $q(u)$, respectively. Figure 4.2 shows that the state x_1 tends to zero, which is within the bound given in (4.29). Clearly, the simulation results verify our theoretical findings and show the effectiveness of our proposed control scheme.

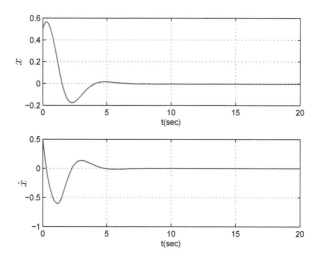

Figure 4.2 Stabilization of x_1 and x_2

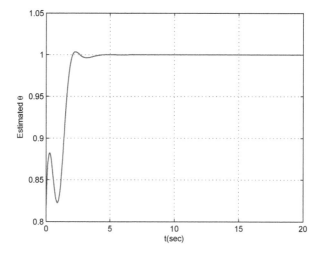

Figure 4.3 Estimation of parameter θ

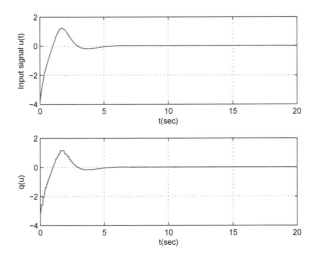

Figure 4.4 Input $u(t)$ **and quantized input** $q(u)$

4.7   Notes

This chapter addresses adaptive backstepping feedback stabilization of a general class of strict-feedback nonlinear systems preceded by quantized input signal. A hysteresis-logarithmic type quantizer is studied. The quantization parameter is chosen based on an established inequality depending on the given controller design parameters and certain system parameters, which ensures system stability and stabilization error within an adjustable bound. The proposed method can be applied to the uniform quantizers, hysteresis-uniform quantizers, log-uniform quantizers, or logarithmic quantizers in Chapter 3.

Acknowledgment

Reprinted from Copyright (2021), with permission from IEEE. Jing Zhou, Changyun Wen and Guanghong Yang, "Adaptive Backstepping Stabilization of Nonlinear Uncertain Systems With Quantized Input Signal", *IEEE Transactions on Automatic Control*, vol. 59, no. 2, pp. 460–464, 2014.

Chapter 5

Adaptive Tracking of Nonlinear Uncertain Systems with Input Quantization

Note that the results in Chapter 4 are only applicable to stabilization of uncertain nonlinear systems with bounded input quantization. Also, the proposed controller in Chapter 4 must follow a guideline to select the parameters of the quantizer. This chapter investigates the tracking problem for a class of uncertain nonlinear systems in the presence of sector-bounded input quantization. The main challenge in solving the tracking problem is how to compensate for the effects of the reference input on the tracking errors. By using the backstepping technique, a new adaptive control algorithm is developed by constructing a new compensation method for the effects of the input quantization. A hyperbolic tangent function is introduced in the controller with a new transformation of the control signal. Unlike the control scheme proposed in Chapter 4, the developed controllers in this chapter do not require the Lipschitz condition for the nonlinear functions and also the quantization parameters can be unknown. Sector-bounded quantizers introduced in Chapter 3 are considered, including the logarithmic quantizer and hysteresis-logarithmic quantizer. Besides showing global stability, tracking error performance is also established and can be adjusted by tuning certain design parameters.

DOI: 10.1201/9781003176626-5   **61**

5.1 Introduction

Quantized control of uncertain systems with known quantization parameters has been studied by using robust approaches in [67, 80] and adaptive approaches in [33,34,101,124,127,142,147]. Due to a number of advantages of backstepping technique, adaptive backstepping control of uncertain nonlinear systems with quantized input has been studied in [124, 127, 142, 147]. However in [142, 147] and Chapter 4, only stabilization problem of uncertain nonlinear systems with input quantization is considered. Although the proposed method can avoid stability conditions depending on the control input, it requires the nonlinear functions to satisfy global Lipschitz conditions with known Lipschitz constants. Also the proposed controller must follow a guideline to select the parameters of the quantizer to ensure stability of the closed-loop system. Such Lipschitz conditions are relaxed in [124] and [145]. In [124], an implicit adaptive controller was developed for the system where unknown parameters only appear in the last differential equation of the system and the controller is contained an equation which is related to a hyperbolic tangent function. Obviously, it is not easy to solve such an equation to get an explicit controller. In [145], a hyperbolic function is introduced into the adaptive controller to compensate for the effects of input quantization. Similarly, a novel smooth function is adopted in [59] to generate the controller which can eliminate the effects of input quantization and actuator faults.

Motivated by its theoretical and practical importance, this chapter is investigating the tracking problem for a class of uncertain nonlinear systems in the presence of input quantization. The main challenge in solving the tracking problem is how to compensate for the effects of the reference input on the tracking errors. By using the backstepping technique, a new adaptive control algorithm is developed by constructing a new compensation method for the effects of the input quantization. A hyperbolic tangent function is introduced in the controller with a new transformation of the control signal. Sector-bounded quantizers in Chapter 3 are considered in the chapter, including logarithmic quantizer and a hysteresis-logarithmic quantizer. Unknown parameters are contained in each differential equation of the system and their bounds are not required to be known. Besides showing global stability, tracking error performance is also established and can be adjusted by tuning certain design parameters. To overcome the difficulty, a new adaptive control scheme is proposed, which has the following features compared the methods presented in Chapter 4.

- The Lipschitz condition required for the nonlinear functions in Chapter 4 is removed.

- Unlike the control scheme proposed in Chapter 4, the proposed controller in this chapter does not need to follow a guideline to guarantee stability.

- The quantization parameters can be unknown.

- By introducing a hyperbolic tangent function, proposing a new transformation of the final control signal, and using the property of the quantizer, the effects from the quantization input are effectively compensated.

■ The developed controller can achieve desired tracking performance for a larger class of nonlinear systems by constructing a new compensation method for the effects of the input quantization.

5.2 Problem Statement

5.2.1 Modeling of Uncertain Nonlinear Systems

A class of uncertain nonlinear systems is considered in the following parametric strict-feedback form as in [53, 71].

$$
\begin{aligned}
\dot{x}_i &= x_{i+1} + \psi_i(\bar{x}_i) + \phi_i^T(\bar{x}_i)\theta \\
\dot{x}_n &= q(u(t)) + \psi_n(\bar{x}_n) + \phi_n^T(\bar{x}_n)\theta \\
y &= x_1(t), \quad i = 1, \ldots, n-1,
\end{aligned} \tag{5.1}
$$

where $x_i(t) \in \Re^1, i = 1, \ldots, n$, $u(t) \in \Re^1$ and $y(t) \in \Re^1$ are the states, input and output of the system respectively, $\bar{x}_i(t) = [x_1(t), \ldots, x_i(t)]^T \in \Re^i$, the vector $\theta \in \Re^r$ is constant and unknown, $\psi_i(\bar{x}_i) \in \Re^1$ and $\phi_i(\bar{x}_i) \in \Re^r$ are known nonlinear functions and differentiable, $q(u(t))$ represents a quantizer and takes the quantized values.

The control objective is to design a feedback control law for $u(t)$ to ensure that the output $y(t)$ can track a reference signal $y_r(t)$ and all closed-loop signals are bounded.

Assumption 5.1 *The reference signal $y_r(t)$ and its nth order derivatives are known and bounded.*

Remark 5.1   The proposed scheme in Chapter 4 and [142, 147] requires the non-linearities in the system to satisfy Lipschitz continuous condition. Compared with them, this condition is now removed. Also in contrast to [124], unknown parameters are contained in each differential equation of the system and their bounds are not required to be known. Thus the system considered in this chapter is more general.

5.2.2 Quantizer

The quantizer $q(u)$ considered in the chapter has the sector bounded property, as shown in Chapter 3.

$$
|q(u) - u| \leq \delta|u| + u_{min}, \tag{5.2}
$$

where $0 < \delta < 1$ and $u_{min} > 0$ are quantization parameters. It is shown in Chapter 3 that a uniform quantizer in (3.1), a logarithmic quantizer in (3.2), and a hysteresis-logarithmic quantizer in (3.4) satisfy the property in (5.2) and can be used in the chapter.

Remark 5.2   The uniform quantizer has the uniformly spaced quantization levels. The logarithmic and hysteresis-logarithmic quantizers have unequal quantization levels. Such kind of quantizers are the coarsest quantizers which minimize the average rate of communication instances and are easy to implement. Compared with the logarithmic quantizer, the hysteresis-logarithmic quantizer has additional quantization levels, which are used to avoid chattering. Detailed discussions can be found in Chapter 3.

5.3   Design of Adaptive Controllers

5.3.1   Control Scheme I for Known Quantization Parameters

In this section, we will design adaptive backstepping feedback control laws for the nonlinear uncertain system (5.2) where the parameters of the quantizers are known. We begin by introducing the change of coordinates

$$
\begin{aligned}
z_1 &= y - y_r & (5.3) \\
z_i &= x_i - \alpha_{i-1}, \; i = 2, 3, \ldots, n, & (5.4)
\end{aligned}
$$

where α_i are virtual controllers. The design procedure is outlined in the following steps.

Step i, ($i = 1, \ldots, n - 1$). The design for the first $n - 1$ subsystems follows the backstepping design procedure in Chapter 2. Design the stabilizing function α_i as

$$
\begin{aligned}
\alpha_i &= -c_i z_i - z_{i-1} - \psi_i - \omega_i^T \hat{\theta} + \sum_{j=1}^{i-1} \frac{\partial \alpha_{i-1}}{\partial x_j} (x_{j+1} + \psi_j) \\
&\quad + \frac{\partial \alpha_{i-1}}{\partial \hat{\theta}} \left(\Gamma \tau_i - \Gamma l_\theta (\hat{\theta} - \theta_0) \right) + \sum_{j=2}^{i-1} \frac{\partial \alpha_{j-1}}{\partial \hat{\theta}} \Gamma \omega_i z_j + y_r^{(i)} & (5.5) \\
\tau_i &= \tau_{i-1} + \omega_i z_i & (5.6) \\
\omega_i &= \phi_i - \sum_{j=1}^{i} \frac{\partial \alpha_{i-1}}{\partial x_j} \phi_j, & (5.7)
\end{aligned}
$$

where c_i and l_θ are positive constants, Γ is a positive definite matrix, $\hat{\theta}$ is an estimate of θ, and θ_0 is a constant vector.

Step n. In the last step n, the actual control input u appears and is at our disposal.

$$u = -\tanh(z_n u_n/\lambda)u_n \tag{5.8}$$

$$u_n = \frac{1}{1-\delta}(-\alpha_n + \mu\tanh(\mu z_n/\lambda)) \tag{5.9}$$

$$\alpha_n = -c_n z_n - z_{n-1} - \psi_n + \sum_{j=1}^{n-1}\frac{\partial\alpha_{n-1}}{\partial x_j}(x_{j+1} + \psi_j) - \omega_n^T\hat{\theta}$$

$$+\frac{\partial\alpha_{n-1}}{\partial\hat{\theta}}\Gamma\left(\tau_n - l_\theta(\hat{\theta} - \theta_0)\right) + \sum_{j=2}^{n-1}\frac{\partial\alpha_{j-1}}{\partial\hat{\theta}}\Gamma\omega_n z_j + y_r^{(n)} \tag{5.10}$$

$$\tau_n = \tau_{n-1} + \omega_n z_n \tag{5.11}$$

$$\omega_n = \phi_n - \sum_{j=1}^{n}\frac{\partial\alpha_{n-1}}{\partial x_j}\phi_j \tag{5.12}$$

$$\dot{\hat{\theta}} = \Gamma\tau_n - \Gamma l_\theta(\hat{\theta} - \theta_0), \tag{5.13}$$

where c_n, λ, and μ are positive constants with $\mu \geq u_{min}$.

Theorem 5.1
Consider the closed-loop adaptive system consisting of plant (5.2) with an input quantization, the adaptive backstepping controller (5.8) with virtual control laws (5.5)–(5.7) and (5.9)–(5.12), parameter estimator with updating law (5.13). The global boundedness of all the signals in the system is ensured. Furthermore, the tracking error $e(t) = y(t) - y_r(t)$ is ultimately bounded as follows:

$$|e(t)| \leq B_1, \text{ where } B_1 = \sqrt{\max\left\{2U_n(0), \frac{2M_1}{F_1}\right\}}, \tag{5.14}$$

where $U_n(0) = \sum_{i=1}^{n}\frac{1}{2}z_i^2(0) + \frac{1}{2}\tilde{\theta}(0)^T\Gamma^{-1}\tilde{\theta}(0)$, $M_1 = 0.557\lambda + \frac{l_\theta}{2}\parallel\theta - \theta_0\parallel^2$, $F_1 = \min\{2c_1, 2c_2, \dots, 2c_n, l_\theta\lambda_{min}(\Gamma)\}$, and $\lambda_{min}(\Gamma)$ is the minimum eigenvalue of Γ.

Proof 5.1 Consider the Lyapunov function as follows

$$U_i = \sum_{j=1}^{i}\frac{1}{2}z_j^2 + \frac{1}{2}\tilde{\theta}^T\Gamma^{-1}\tilde{\theta}, \; i = 1, 2, \dots, n. \tag{5.15}$$

where $\tilde{\theta} = \theta - \hat{\theta}$. The derivative of U_n satisfies

$$\dot{U}_n = \dot{U}_{n-1} + z_n\left(q(u) + \psi_n + \theta^T\phi_n - \frac{\partial\alpha_{n-1}}{\partial\hat{\theta}}\dot{\hat{\theta}} - y_r^{(n)}\right.$$

$$\left. - \sum_{j=1}^{n-1}\frac{\partial\alpha_{n-1}}{\partial x_j}(x_{j+1} + \psi_j + \theta^T\phi_j)\right). \tag{5.16}$$

The following inequality is derived by multiplying $|z_n|$ on both sides of (5.2) and using (5.8)

$$
\begin{aligned}
z_n q(u) &\leq z_n u + \delta |z_n u| + u_{min} |z_n| \\
&\leq -z_n u_n \tanh(z_n u_n / \lambda) + \delta |z_n u_n \tanh(z_n u_n / \lambda)| + u_{min} |z_n| \\
&\leq -(1-\delta) z_n u_n \tanh(z_n u_n / \lambda) + u_{min} |z_n| \\
&\leq -(1-\delta)|z_n u_n| + u_{min} |z_n| + \epsilon_1 \\
&\leq -(1-\delta) z_n u_n + \mu |z_n| + \epsilon_1,
\end{aligned}
\tag{5.17}
$$

where $\epsilon_1 = 0.2785\lambda(1-\delta)$ and we have used the property that $|x| - x\tanh(x/\lambda) \leq 0.2785\lambda$ in [83, 118]. Using (5.9), (5.10), (5.13), (5.16), and (5.17), the derivative of U_n satisfies

$$
\begin{aligned}
\dot{U}_n &\leq \dot{U}_{n-1} + z_n\left(\alpha_n + \psi_n + \theta^T \phi_n - \frac{\partial \alpha_{n-1}}{\partial \hat{\theta}}\dot{\hat{\theta}} - y_r^{(n)}\right. \\
&\quad \left. - \sum_{j=1}^{n-1} \frac{\partial \alpha_{n-1}}{\partial x_j}(x_{j+1} + \psi_j + \theta^T \phi_j)\right) - \mu z_n \tanh(\mu z_n / \lambda) + \mu |z_n| + \epsilon_1 \\
&\leq -\sum_{j=1}^n c_j z_j^2 + \tilde{\theta}^T\left(\tau_n - \Gamma^{-1}\dot{\hat{\theta}}\right) \\
&\quad + \left(\sum_{j=1}^{n-1} \frac{\partial \alpha_j}{\partial \hat{\theta}} z_{j+1}\right)\left(\Gamma \tau_n - \Gamma l_\theta(\hat{\theta} - \theta_0) - \dot{\hat{\theta}}\right) + \epsilon_2 \\
&\leq -\sum_{i=1}^n c_i z_i^2 - \frac{1}{2} l_\theta \| \tilde{\theta} \|^2 + \epsilon_2 + \frac{1}{2} l_\theta \| \theta - \theta_0 \|^2 \\
&\leq -F_1 U_n + M_1,
\end{aligned}
\tag{5.18}
$$

where $\epsilon_2 = 0.2785\lambda(2 - \delta) \leq 0.557\lambda$ and the following property is used.

$$
l_\theta \tilde{\theta}(\hat{\theta} - \theta_0) \leq -\frac{1}{2} l_\theta \| \tilde{\theta} \|^2 + \frac{1}{2} l_\theta \| (\theta - \theta_0) \|^2 .
\tag{5.19}
$$

By direct integration of the differential inequality (5.18), we have

$$
U_n \leq U_n(0)e^{-F_1 t} + \frac{M_1}{F_1}(1 - e^{-F_1 t}),
\tag{5.20}
$$

which shows that U_n is uniformly bounded, yielding that $z_1, z_2, \ldots, z_n$ and $\hat{\theta}$ are all bounded. The boundedness of x_i ($i = 1, \ldots, n$) can be ensured from the boundedness of α_i in (5.5) and the n-th order derivatives of y_r, and the fact that $x_i = z_i + \alpha_{i-1}$ and $x_1 = z_1 + y_r$. Combining this with (5.8) and (5.9), $u(t)$ is bounded. Thus all the signals of the overall closed-loop system are globally uniformly bounded. Particularly, the bound of z_i is bounded in the set $\left\{z \mid \|z\| \leq \sqrt{\max\left\{2U_n(0), \frac{2M_1}{F_1}\right\}}\right\}$, which is adjustable by tuning the design parameters c_i, l_θ θ_0, and $\lambda_{min}(\Gamma)$.

Remark 5.3 The controller designed in this section achieves the goals of stabilization and tracking with quantized input signal. The difficulty to achieve the control objective is to handle the quantization error because its bound depends on the control input $u(t)$. In [142, 147], global Lipschitz condition for the nonlinear functions is required to guarantee the stability and compensate for the effects of the quantization error. In this chapter, a new controller is developed in (5.8) which is a function of the virtual controller u_n and includes a hyperbolic tangent function $\tanh(z_n u_n / \lambda)$. Together with the property of the quantizer, this new control strategy enables the effects from the quantization error $|\delta z_n u|$ to be compensated by taking out $z_n u$ from the absolute function and transforming it to a term including only the virtual control signal $u_n(t)$ as shown in (5.17). As a result, the global Lipschitz continuous restriction in [142, 147] for the nonlinear functions is removed, which largely broadens the class of systems to be controlled. In [124], the control signal is implicitly involved in the proposed control law to compensate for the effects of input quantization. Compared to [124], the proposed new control signal is an explicit function of the states and estimated parameters, and thus easy for implementation in practice. In addition, unknown parameters are contained in each differential equation of the system considered in our chapter and their bounds are not required to be known.

Remark 5.4 The inequality (5.17) is a key step. It transforms the quantized input term $z_n q(u)$ to $-(1 - \delta)z_n u_n$ which is an explicit function of the virtual control signal u_n and can be directly designed based on Lyapunov stability.

Remark 5.5 The ultimate tracking error is adjustable and can be made smaller by increasing the design parameter c_i and $\lambda_{min}(\Gamma)$. Note that λ used in the hyperbolic tangent function in (5.8) should be chosen as a suitable positive constant. While the tracking error becomes theoretically small for sufficiently small λ, the tangent hyperbolic function approaches the sign function. Thus, there is a trade-off between the tracking performance and the realization of controller.

5.3.2 Control Scheme II for Unknown Quantization Parameters

In this section, we consider the case that the parameters δ and u_{min} of the quantizer are unknown. So far there is no result available to address this issue due to the challenge of the problem involved. It is also difficult to find a feasible solution to the adaptive control problem formulated if we design estimators to directly identify these parameters. After extensive research, an innovative solution is arrived at by proposing suitable estimators to identify the bounds of certain parameters related to these unknown quantized parameters. As the first $n - 1$ steps in the recursive adaptive backstepping process are the same as the design procedure in previous section, we only focus on the last step, which gives the control input u and the estimators to identify the unknown parameters summarized below.

$$u = -\tanh(z_n u_n/\lambda)u_n \tag{5.21}$$

$$u_n = \hat{\beta}(-\alpha_n + \hat{\mu}\tanh(z_n/\lambda)) \tag{5.22}$$

$$\dot{\hat{\theta}} = \Gamma\tau_n - \Gamma l_\theta(\hat{\theta} - \theta_0) \tag{5.23}$$

$$\dot{\hat{\mu}} = \gamma_1 z_n \tanh(z_n/\lambda) - \gamma_1 l_1(\hat{\mu} - \mu_0) \tag{5.24}$$

$$\dot{\hat{\beta}} = \gamma_2 z_n(-\alpha_n + \hat{\mu}\tanh(z_n/\lambda)) - \gamma_2 l_2(\hat{\beta} - \beta_0), \tag{5.25}$$

where γ_1, l_1, γ_2, l_2, μ_0, and β_0, are positive constants, $\hat{\mu}$ and $\hat{\beta}$ are estimates of the bound $\mu \geq u_{min}$ and $\beta = \frac{1}{1-\delta}$.

Theorem 5.2

Consider the closed-loop adaptive system consisting of plant (5.2) with an input quantization, the adaptive backstepping controller (5.21) with virtual control laws (5.5)–(5.7), (5.10), (5.12), and (5.22), the parameter estimators with updating laws (5.23), (5.24), and (5.25). The global boundedness of all the signals in the system is ensured. Furthermore, the tracking error $e(t) = y(t) - y_r(t)$ is ultimately bounded as follows:

$$|e(t)| \leq B_2, \text{ where } B_2 = \sqrt{\max\left\{2U_n(0), \frac{2M_2}{F_2}\right\}}, \tag{5.26}$$

where $F_2 = \min\{2c_1, 2c_2, \ldots, 2c_n, l_\theta\lambda_{min}(\Gamma), l_1\gamma_1, l_2\gamma_2\}$, $M_2 = 0.557\lambda + \frac{l_\theta}{2}$ $\|\theta - \theta_0\|^2 + \frac{l_1}{2}(\mu - \mu_0)^2 + \frac{l_2}{2}(1 - \delta)(\beta - \beta_0)^2$, $U_n(0) = \sum_{i=1}^{n}\frac{1}{2}z_i(0)^2 + \frac{1}{2}\tilde{\theta}^T(0)\Gamma^{-1}\tilde{\theta}(0) + \frac{1}{2\gamma_1}\tilde{\mu}^2(0) + \frac{(1-\delta)}{2\gamma_2}\tilde{\beta}^2(0)$.

Proof 5.2 We choose the final Lyapunov function as follows

$$U_n = \sum_{i=1}^{n}\frac{1}{2}z_i^2 + \frac{1}{2}\tilde{\theta}^T\Gamma^{-1}\tilde{\theta} + \frac{1}{2\gamma_1}\tilde{\mu}^2 + \frac{(1-\delta)}{2\gamma_2}\tilde{\beta}^2. \tag{5.27}$$

Now substituting (5.21) and (5.22) into (5.17), the following inequality is obtained.

$$\begin{aligned}
z_n q(u) &\leq -(1-\delta)z_n u_n + \mu|z_n| + \epsilon_1 - (1-\delta)(\beta - \hat{\beta})z_n \bar{u}_n + \mu|z_n| + \epsilon_1 \\
&= -z_n \bar{u}_n + (1-\delta)\hat{\beta}z_n \bar{u}_n + \mu|z_n| + \epsilon_1 \\
&\leq z_n \alpha_n + (1-\delta)\tilde{\beta}z_n \bar{u}_n - \hat{\mu}z_n \tanh(z_n/\lambda) + \mu z_n \tanh(z_n/\lambda) + \epsilon_2 \\
&\leq z_n \alpha_n + (1-\delta)\tilde{\beta}z_n \bar{u}_n + \tilde{\mu}z_n \tanh(\frac{z_n}{\lambda}) + \epsilon_2,
\end{aligned} \tag{5.28}$$

where $\epsilon_2 = 0.2785\lambda(1 - \delta + \mu)$ and $\bar{u}_n = -\alpha_n + \hat{\mu}\tanh(z_n/\lambda)$. Using (5.24),

(5.25), and (5.28), the derivative of U_n satisfies

$$
\begin{aligned}
\dot{U}_n &\leq -\sum_{i=1}^{n} c_i z_i^2 - \frac{1}{2} l_\theta \parallel \tilde{\theta} \parallel^2 + \frac{1}{2} l_\theta \parallel \theta - \theta_0 \parallel^2 + \epsilon_2 \\
&\quad + \frac{1}{\gamma_1} \tilde{\mu} \left(\gamma_1 z_n \tanh(z_n/\lambda) - \dot{\hat{\mu}} \right) + \frac{(1-\delta)}{\gamma_2} \tilde{\beta} \left(\gamma_2 z_n \bar{u}_n - \dot{\hat{\beta}} \right) \\
&\leq -\sum_{i=1}^{n} c_i z_i^2 - \frac{l_\theta}{2} \parallel \tilde{\theta} \parallel^2 - \frac{l_1}{2} \tilde{\mu}^2 - \frac{l_2(1-\delta)}{2} \tilde{\beta}^2 + M_2 \\
&\leq -F_2 U_n + M_2.
\end{aligned}
\tag{5.29}
$$

By direct integration of the differential inequality (5.29), we have

$$
U_n \leq U_n(0) e^{-F_2 t} + \frac{M_2}{F_2} (1 - e^{-F_2 t}).
\tag{5.30}
$$

Based on (5.30), all signals of the overall closed-loop system are globally uniformly bounded and $z_i(t)$ approaches to a compact set $\{z_i \mid |z_i(t)| \leq B_2\}$ where B_2 is given in (5.26).

Remark 5.6 The virtual control laws $\alpha_i (i = 1, \dots, n)$ are the same for both cases of known quantization parameters and unknown quantization parameters. When the parameters δ and u_{min} of the quantizer are unknown, two on-line estimators (5.24) and (5.25) are developed and the estimates $\hat{\mu}$ and $\hat{\beta}$ are used in the adaptive controller (5.22). Note that, instead of directly estimating the unknown quantization parameters u_{min} and δ, estimators (5.24) and (5.25) are designed to identify two parameters related to them.

5.4   Simulation Results

In this section we consider an uncertain nonlinear system with quantization input as follows.

$$
\ddot{x} + \theta \sin(\dot{x}) + x^2 = q(u),
\tag{5.31}
$$

where θ is an unknown parameter and $q(u)$ is a quantized input. The objective is to design a quantized control input for u to make the output $y = x$ track the reference signal $y_r(t) = \sin(t)$. In the simulation, we consider three quantizers: uniform, logarithmic and hysteresis-logarithmic . The actual parameter value is chosen as $\theta = 1$ for simulation.

■ **Case 1:** Known quantization parameters.
The quantization parameters are chosen as $l = 0.2$, $\delta = 0.2$, and $u_{min} = 0.1$. The initial states and parameter are set as $x(0) = 0.5$, $\dot{x}(0) = 0.2$ and $\hat{\theta}(0) = 0.8$. The control design parameters are chosen as $c_1 = c_2 = 6$, $\Gamma = 1$, $l_\theta =$

$0.01, \theta_0 = 0, \lambda = 0.2$, and $\mu = 0.1$. The trajectory output, tracking error, the control signal and the quantized control are shown in Figure 5.1 for a uniform quantizer, Figure 5.2 for a logarithmic quantizer, and Figure 5.3 for a hysteresis-logarithmic quantizer, respectively. For three input quantizers, the simulation results show that the output tracks the desired reference signal and the tracking error is bounded. The simulation results in Figures 5.4–5.6 show that the magnitudes of tracking errors with control parameters $c_1 = c_2 = 1$ are larger than those with parameters $c_1 = c_2 = 6$ in Figures 5.1–5.3. This also verifies our theoretical findings in Theorem 5.1 that the tracking error can be made smaller by increasing c_i.

■ **Case 2:** Unknown quantization parameter.
When the quantized parameters are unknown, the adaptive backstepping controller (5.21)–(5.25) are employed. The initial states and parameter are set as $x(0) = 0, \dot{x}(0) = 0.9$ and the control parameters are chosen as $c_1 = 8, c_2 = 5, \Gamma = 1, \gamma_2 = 0.01$. Figures 5.7, 5.8, and 5.9 respectively show the trajectories of output, tracking error and the control signal for the system (5.31) preceded by three quantizers with unknown quantization parameters. The simulation results for three quantizers verify our theoretical findings in Theorem 5.2 that the output tracks the desired reference signal and the tracking errors are bounded.

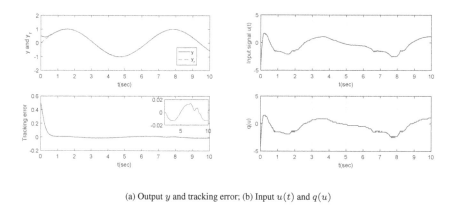

(a) Output y and tracking error; (b) Input $u(t)$ and $q(u)$

Figure 5.1 Uniform quantizer with $c_1 = c_2 = 6$

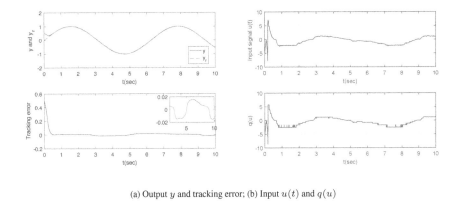

(a) Output y and tracking error; (b) Input $u(t)$ and $q(u)$

Figure 5.2 Logarithmic quantizer with $c_1 = c_2 = 6$

5.5 Notes

In this chapter, we propose an adaptive backstepping approaches for single-loop uncertain nonlinear systems with input sector-bounded quantization. By introducing a hyperbolic tangent function, proposing a new transformation of the final control signal and using the property of the quantizer, the effects from the quantization input are effectively compensated and the global Lipschitz conditions required for the nonlinearities are relaxed. When quantized parameters are not known, new parameter updating laws are developed which do not require the knowledge on the bounds of such unknown parameters. Besides showing global stability of the system, the tracking error can asymptotically converge to a residual, which can be made smaller by choosing suitable design parameters and thus adjustable. It is noted that the results received in this chapter can be directly applied to bounded input quantizers.

Acknowledgment

Reprinted from Copyright (2021), with permission from Elsevier. Jing Zhou, Changyun Wen and Wei Wang, "Adaptive control of uncertain nonlinear systems with quantized input signal", *Automatica*, vol. 95, pp. 152–162, 2018.

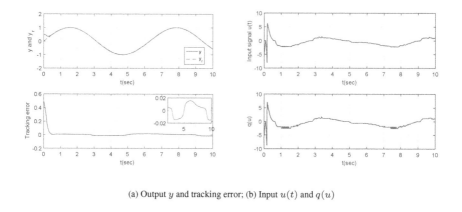

(a) Output y and tracking error; (b) Input $u(t)$ and $q(u)$

Figure 5.3 Hysteresis-logarithmic quantizer with $c_1 = c_2 = 6$

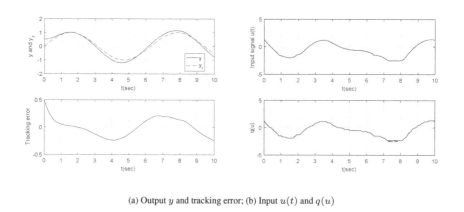

(a) Output y and tracking error; (b) Input $u(t)$ and $q(u)$

Figure 5.4 Uniform quantizer with $c_1 = c_2 = 1$

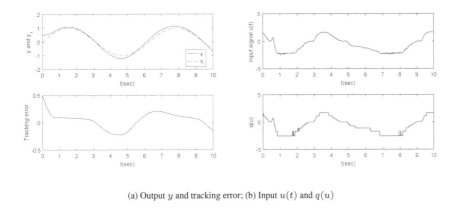

(a) Output y and tracking error; (b) Input $u(t)$ and $q(u)$

Figure 5.5 Logarithmic quantizer with $c_1 = c_2 = 1$

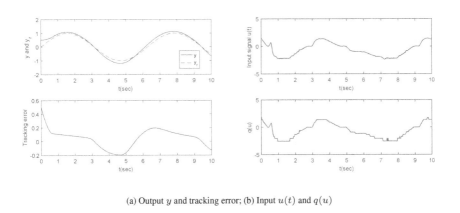

(a) Output y and tracking error; (b) Input $u(t)$ and $q(u)$

Figure 5.6 Hysteresis-logarithmic quantizer with $c_1 = c_2 = 1$

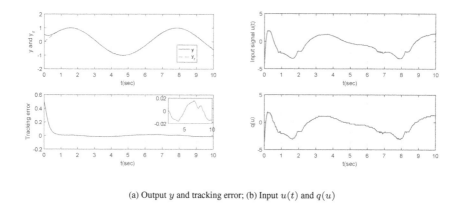

(a) Output y and tracking error; (b) Input $u(t)$ and $q(u)$

Figure 5.7 Uniform quantizer with unknown parameters

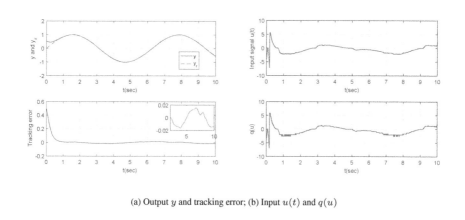

(a) Output y and tracking error; (b) Input $u(t)$ and $q(u)$

Figure 5.8 Logarithmic quantizer with unknown parameters

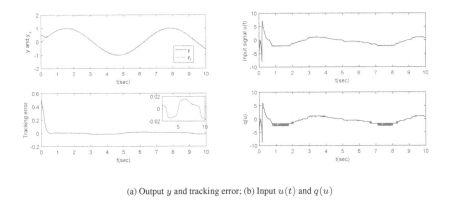

(a) Output y and tracking error; (b) Input $u(t)$ and $q(u)$

Figure 5.9 Hysteresis-logarithmic quantizer with unknown parameters

Chapter 6

Decentralized Adaptive Control of Interconnected Systems with Input Quantization

Due to the difficulties in considering the effects of both uncertain interconnections and quantization input, the extension of single-loop results to multi-loop interconnected systems is challenging. In this chapter, we extend our approach in Chapter 5 to control a class of nonlinear interconnected systems in the presence of input quantization. A totally decentralized adaptive control scheme is developed with a new compensation method incorporated for the unknown nonlinear interactions and quantization error. Each local controller, designed simply based on the model of each subsystem by using the adaptive backstepping technique, only employs local information to generate control signals. Besides showing stability, tracking error performance is also established and can be adjusted by tuning certain design parameters.

6.1 Introduction

Due to the difficulties in considering the effects of uncertain interconnections, the extension of single-loop results to multi-loop interconnected systems is challenging, especially since both the input quantization and unknown interconnections are considered. In the control of uncertain interconnected systems, decentralized adaptive

DOI: 10.1201/9781003176626-6

control strategy, designed independently for local subsystems and using only locally available signals for feedback propose, is an efficient and practical strategy, see for examples [38, 109]. Research on decentralized adaptive control using backstepping technique has also received great attention, see for examples [48, 108, 117, 119]. In the presence of input quantization in unknown interconnected systems, the result of available decentralized control is still limited. [101] has addressed the issue of decentralized quantized control via output-feedback for interconnected systems. In [101], the original system needs to be transformed to a form including only the output signal and the signals from filters. So interactions only exist in the equation for the output in the final control systems and the rest equations related to the filter signals do not involve interactions. In this chapter, a more general class of interconnected systems is considered in the sense that interactions exist in all the differential equations of the subsystems. Thus for such systems, it is more challenging to design appropriate controllers to account for the effects of unknown interactions.

In this chapter, we propose adaptive backstepping approaches to solve the tracking control problem of multi-loop uncertain interconnected nonlinear systems, which are preceded by the quantized input signal. The control signal is quantized by a class of sector-bounded quantizers including the uniform quantizer, the logarithmic quantizer, and the hysteresis quantizer. Unknown parameters are contained in each differential equation of the system and their bounds are not required to be known. Considering multi-loop interconnected systems with interactions allowed in every state equation, a totally decentralized adaptive controller design approach is developed together with a new compensation method constructed for the unknown nonlinear interactions and quantization error. A well-defined smooth function is introduced into the decentralized adaptive controllers to compensate for the effects of unknown nonlinear interactions. Besides showing the global stability of the systems, the tracking error can asymptotically converge to a residual, which can be made arbitrarily small by choosing suitable design parameters.

6.2 Problem Formulation

A class of interconnected systems consisting N single-input and single-output subsystems is considered in the following.

$$\dot{x}_{i,j} = x_{i,j+1} + \psi_{i,j}(\bar{x}_{i,j}) + \phi_{i,j}^T(\bar{x}_{i,j})\theta_i + h_{i,j}(y_1, \ldots, y_N) \quad (6.1)$$
$$\dot{x}_{i,n_i} = q_i(u_i) + \psi_{i,n_i}(\bar{x}_{i,n_i}) + \phi_{i,n_i}^T(\bar{x}_{i,n_i})\theta_i + h_{i,n_i}(y_1, \ldots, y_N)$$
$$y_i(t) = x_{i,1}(t), \quad i = 1, \ldots, N, \quad j = 1, \ldots, n_i - 1 \quad (6.2)$$

where $x_{i,j}(t) \in \Re^1$, $u_i(t) \in \Re^1$ and $y_i(t) \in \Re^1$ $i = 1, \ldots, N$, $j = 1, \ldots, n_i$, are the state, input and output of the subsystem respectively, $\bar{x}_{i,j}(t) = [x_{i,1}(t), \ldots, x_{i,j}(t)]^T \in \Re^j$ the vector $\theta_i \in \Re^{r_i}$ is constant and unknown, $\psi_{i,j} \in \Re^1$ and $\phi_{i,j}(.) \in \Re^{r_i}$ are known smooth nonlinear functions, $h_{i,j}(.)$ denotes the nonlinear interaction from the jth subsystem to the ith subsystem for $j \neq i$ or a nonlinear unmodeled part of the ith subsystem for $j = i$, the input $q_i(u_i)$ represents a quantizer.

For such a class of systems, the following assumptions are made.

Assumption 6.1 *The nonlinear interactions satisfy*

$$\left(h_{i,j}(y_1,\ldots,y_N,t)\right)^2 \le \sum_{k=1}^{N} r_{i,j,k}\bar{h}_{i,j,k}(y_k), \tag{6.3}$$

where $\bar{h}_{i,j,k}(.)$ *are known smooth functions and* $r_{i,j,k}$ *are positive constants denoting the strength of the uncertain subsystem interactions.*

Assumption 6.2 *The reference signal* $y_{ri}(t)$ *and its* n_i*-th order derivatives are known and bounded.*

The control objective is to design a totally decentralized adaptive controller for system (6.2) such that the closed-loop system is stable and the output $y_i(t)$ can track a given reference signal $y_{ri}(t)$ as close as possible.

The quantizer $q(u)$ considered in the chapter has the sector-bounded property as in (3.7).

$$|q(u) - u| \le \delta|u| + u_{min}, \tag{6.4}$$

where $0 \le \delta < 1$ and $u_{min} > 0$ are quantization parameters. It is shown in Chapter 3 that the logarithmic quantizer in (3.2) and the hysteresis-logarithmic quantizer in (3.4) satisfy the property in (6.4). It is note that uniform quantizer in (3.1), hysteresis-uniform quantizer in (3.3), and logarithmic-uniform quantizer in (3.5) also satisfy the property in (6.4) with $\delta = 0$. Above mentioned quantizers can be used in the chapter. The mathematical descriptions and discussions of above quantizers are given in Chapter 3.

6.3 Design of Decentralized Adaptive Controller

As usual in the backstepping design, the following change of coordinates is made.

$$z_{i,1} = y_i - y_{ri} \tag{6.5}$$

$$z_{i,j} = x_{i,j} - \alpha_{i,j-1},\ j = 2, 3, \ldots, n_i \tag{6.6}$$

where $\alpha_{i,j-1}$ are virtual controllers.

The design procedure is elaborated in the following steps.

Step 1. We start with the first equation of (6.2). The derivative of tracking error $z_{i,1}$ is given as

$$\begin{aligned}
\dot{z}_{i,1}(t) &= x_{i,2}(t) + \psi_{i,1}(\bar{x}_{i,1}) + \phi_{i,1}^T(\bar{x}_{i,1})\theta_i + h_{i,1}(y_1,..,y_N) - \dot{y}_{ri} \\
&= z_{i,2} + \alpha_{i,1} + \psi_{i,1} + \phi_{i,1}^T\theta_i + h_{i,1}(y_1,..,y_N) - \dot{y}_{ri} \tag{6.7}
\end{aligned}$$

The virtual control law $\alpha_{i,1}$ is designed as

$$
\begin{aligned}
\alpha_{i,1} &= -c_{i,1}z_{i,1} - \frac{1}{4}z_{i,1} - \psi_{i,1}(\bar{x}_{i,1}) - \phi_{i,1}^T(\bar{x}_{i,1})\hat{\theta}_i + \dot{y}_{ri} \\
&\quad -sg_i(z_{i,1}) \sum_{j=1}^{n_i}(n_i - j + 1) \sum_{k=1}^{N} r_{i,j,k}\bar{h}_{k,j,i}(y_i)
\end{aligned}
\tag{6.8}
$$

where $c_{i,1}$ is a positive constant, and $\hat{\theta}_i$ is an estimate of θ_i. The we have

$$
\begin{aligned}
\dot{z}_{i,1}(t) &= z_{i,2} - c_{i,1}z_{i,1} - \frac{1}{4}z_{i,1} + \phi_{i,1}^T\tilde{\theta}_i + h_{i,1}(y_1,..,y_N) \\
&\quad -sg_i(z_{i,1}) \sum_{j=1}^{n_i}(n_i - j + 1) \sum_{k=1}^{N} r_{i,j,k}\bar{h}_{k,j,i}(y_i)
\end{aligned}
\tag{6.9}
$$

where $\tilde{\theta}_i = \theta_i - \hat{\theta}_i$.

The local Lyapunov function is chosen as

$$
U_{i,1} = \frac{1}{2}z_{i,1}^2 + \frac{1}{2}\tilde{\theta}_i^T\Gamma_i^{-1}\tilde{\theta}_i,
\tag{6.10}
$$

where Γ_i is a positive definite matrix. The derivative of $U_{i,1}$ is given by

$$
\begin{aligned}
\dot{U}_{i,1} &\leq z_{i,1}z_{i,2} - c_{i,1}z_{i,1}^2 + \phi_{i,1}^T\tilde{\theta}_i z_{i,1} - \tilde{\theta}_i^T\Gamma_i^{-1}\dot{\hat{\theta}}_i + h_{i,1}(y_1,\ldots,y_N)z_{i,1} \\
&\quad -\frac{1}{4}z_{i,1}^2 - z_{i,1}sg_i(z_{i,1}) \sum_{j=1}^{n_i}(n_i - j + 1) \sum_{k=1}^{N} r_{i,j,k}\bar{h}_{k,j,i}(y_i)
\end{aligned}
\tag{6.11}
$$

Using young's inequality and (6.3) in Assumption 6.1, we have

$$
\begin{aligned}
-\frac{1}{4}z_{i,1}^2 + h_{i,1}(y_1,..,y_N)z_{i,1} &\leq (h_{i,1}(y_1,\ldots,y_N))^2 \\
&\leq \sum_{k=1}^{N} r_{i,1,k}\bar{h}_{i,1,k}(y_k).
\end{aligned}
\tag{6.12}
$$

Define a tuning function $\tau_{i,1}$ as follows.

$$
\tau_{i,1} = \phi_{i,1}z_{i,1}
\tag{6.13}
$$

It follows that

$$
\begin{aligned}
\dot{U}_{i,1} &\leq z_{i,1}z_{i,2} - c_{i,1}z_{i,1}^2 - \tilde{\theta}_i^T\left(\Gamma_i^{-1}\dot{\hat{\theta}}_i - \tau_{i,1}\right) + \sum_{k=1}^{N} r_{i,1,k}\bar{h}_{i,1,k}(y_k) \\
&\quad -z_{i,1}sg_i(z_{i,1}) \sum_{j=1}^{n_i}(n_i - j + 1) \sum_{k=1}^{N} r_{i,j,k}\bar{h}_{k,j,i}(y_i)
\end{aligned}
\tag{6.14}
$$

Remark 6.1   The main difficulty is to handle the unknown interaction $h_{i,1}$ in the differential equation. Compared with the adaptive control strategy designed for single-loop nonlinear systems with input quantization, the new term $-sg_i(z_{i,1}) \sum_{j=1}^{n_i} (n_i - j + 1) \sum_{k=1}^{N} r_{i,j,k} \bar{h}_{k,j,i}(y_i)$ is introduced in the local control law $\alpha_{i,1}$ in (6.8) to compensate for the effects of interactions $h_{i,j}$ among all subsystems $i = 1, 2, \ldots, n_i$.

Step 2. We start with the second tracking error for $z_{i,2}$.

$$
\begin{aligned}
\dot{z}_{i,2}(t) &= \dot{x}_{i,2} - \dot{\alpha}_{i,1} \\
&= x_{i,3} + \psi_{i,2} + \phi_{i,2}^T \theta_i + h_{i,2}(y_1, \ldots, y_N) - \frac{\partial \alpha_{i,1}}{\partial \hat{\theta}_i} \dot{\hat{\theta}}_i \\
&\quad - \frac{\partial \alpha_{i,1}}{\partial x_{i,1}} (x_{i,2} + \psi_{i,1} + \phi_{i,1}^T \theta_i + h_{i,1}(y_1, \ldots, y_N)) - \ddot{y}_{ri} \\
&= z_{i,3} + \alpha_{i,2} + \psi_{i,2} + \phi_{i,2}^T \theta_i + h_{i,2}(y_1, \ldots, y_N) - \frac{\partial \alpha_{i,1}}{\partial \hat{\theta}_i} \dot{\hat{\theta}}_i \\
&\quad - \frac{\partial \alpha_{i,1}}{\partial x_{i,1}} (x_{i,2} + \psi_{i,1} + \phi_{i,1}^T \theta_i + h_{i,1}(y_1, \ldots, y_N)) - \ddot{y}_{ri} \quad (6.15)
\end{aligned}
$$

The virtual control law $\alpha_{i,2}$ is designed as

$$
\begin{aligned}
\alpha_{i,2} &= -c_{i,2} z_{i,2} - \frac{1}{4} z_{i,2} - \frac{1}{4} \left(\frac{\partial \alpha_{i,1}}{\partial x_{i,1}} \right)^2 z_{i,2} - z_{i,1} + \ddot{y}_{ri} \\
&\quad - \psi_{i,2} + \frac{\partial \alpha_{i,1}}{\partial x_{i,1}} (x_{i,2} + \psi_{i,1}) - \hat{\theta}_i^T \left(\phi_{i,2} - \frac{\partial \alpha_{i,1}}{\partial x_{i,1}} \phi_{i,1}^T \right) \\
&\quad + \frac{\partial \alpha_{i,1}}{\partial \hat{\theta}_i} \left(\Gamma_i \tau_{i,2} - \Gamma_i l_{\theta i}(\hat{\theta}_i - \theta_{i0}) \right) \quad (6.16)
\end{aligned}
$$

where $c_{i,2}$ is a positive constant. Then the derivation of $z_{i,2}$ is given as

$$
\begin{aligned}
\dot{z}_{i,2} &= z_{i,3} - c_{i,2} z_{i,2} - \frac{1}{4} z_{i,2} - \frac{1}{4} \left(\frac{\partial \alpha_{i,1}}{\partial x_{i,1}} \right)^2 z_{i,2} - z_{i,1} - \frac{\partial \alpha_{i,1}}{\partial \hat{\theta}_i} \dot{\hat{\theta}}_i \\
&\quad + \tilde{\theta}_i^T \left(\phi_{i,2} - \frac{\partial \alpha_{i,1}}{\partial x_{i,1}} \phi_{i,1} \right) + \frac{\partial \alpha_{i,1}}{\partial \hat{\theta}_i} \left(\Gamma_i \tau_{i,2} - \Gamma_i l_{\theta i}(\hat{\theta}_i - \theta_{i0}) \right) \\
&\quad + h_{i,2}(y_1, \ldots, y_N) - \frac{\partial \alpha_{i,1}}{\partial x_{i,1}} h_{i,1}(y_1, \ldots, y_N) \quad (6.17)
\end{aligned}
$$

The local Lyapunov function is chosen as

$$
U_{i,2} = U_{i,1} + \frac{1}{2} z_{i,2}^2, \quad (6.18)
$$

The derivative of $U_{i,2}$ is given by

$$
\begin{aligned}
\dot{U}_{i,2} \leq{} & z_{i,2}z_{i,3} - c_{i,1}z_{i,1}^2 - c_{i,2}z_{i,2}^2 - \tilde{\theta}_i^T\left(\Gamma_i^{-1}\dot{\hat{\theta}}_i - \tau_{i,1}\right) + \sum_{k=1}^{N} r_{i,1,k}\bar{h}_{i,1,k}(y_k) \\
& +\tilde{\theta}_i^T\left(\phi_{i,2} - \frac{\partial\alpha_{i,1}}{\partial x_{i,1}}\phi_{i,1}\right)z_{i,2} - \frac{1}{4}z_{i,2}^2 - \frac{1}{4}\left(\frac{\partial\alpha_{i,1}}{\partial x_{i,1}}\right)^2 z_{i,2}^2 \\
& +z_{i,2}\frac{\partial\alpha_{i,1}}{\partial\hat{\theta}_i}\left(\Gamma_i\tau_{i,2} - \Gamma_i l_{\theta i}(\hat{\theta}_i - \theta_{i0}) - \dot{\hat{\theta}}_i\right) \\
& +z_{i,2}\left(h_{i,2}(y_1,\ldots,y_N) - \frac{\partial\alpha_{i,1}}{\partial x_{i,1}}h_{i,1}(y_1,\ldots,y_N)\right) \\
& -z_{i,1}sg_i(z_{i,1})\sum_{j=1}^{n_i}(n_i - j + 1)\sum_{k=1}^{N} r_{i,j,k}\bar{h}_{k,j,i}(y_i) \qquad (6.19)
\end{aligned}
$$

Define the second tuning function $\tau_{i,2}$ as

$$
\tau_{i,2} = \tau_{i,1} + \omega_{i,1}z_{i,2} \qquad (6.20)
$$
$$
\omega_{i,2} = \phi_{i,2} - \frac{\partial\alpha_{i,1}}{\partial x_{i,1}}\phi_{i,1} \qquad (6.21)
$$

Similar to Step 1, using young's inequality and (6.3), we have

$$
\begin{aligned}
& z_{i,2}\left(h_{i,2}(y_1,\ldots,y_N) + \frac{\partial\alpha_{i,1}}{\partial x_{i,1}}h_{i,1}(y_1,\ldots,y_N)\right) - \frac{1}{4}z_{i,2}^2 - \frac{1}{4}\left(\frac{\partial\alpha_{i,1}}{\partial x_{i,1}}\right)^2 z_{i,2}^2 \\
& \leq \sum_{j=1}^{2}(h_{i,j}(y_1,\ldots,y_N))^2 \leq \sum_{j=1}^{2}\sum_{k=1}^{N} r_{i,j,k}\bar{h}_{i,j,k}(y_k) \qquad (6.22)
\end{aligned}
$$

It follows that

$$
\begin{aligned}
\dot{U}_{i,2} \leq{} & z_{i,2}z_{i,3} - c_{i,1}z_{i,1}^2 - c_{i,2}z_{i,2}^2 + z_{i,2}\frac{\partial\alpha_{i,1}}{\partial\hat{\theta}_i}\left(\Gamma_i\tau_{i,2} - \Gamma_i l_{\theta i}(\hat{\theta}_i - \theta_{i0}) - \dot{\hat{\theta}}_i\right) \\
& -\tilde{\theta}_i^T\left(\Gamma_i^{-1}\dot{\hat{\theta}}_i - \tau_{i,2}\right) + \sum_{k=1}^{N} r_{i,1,k}\bar{h}_{i,1,k}(y_k) + \sum_{j=1}^{2}\sum_{k=1}^{N} r_{i,j,k}\bar{h}_{i,j,k}(y_k) \\
& -z_{i,1}sg_i(z_{i,1})\sum_{j=1}^{n_i}(n_i - j + 1)\sum_{k=1}^{N} r_{i,j,k}\bar{h}_{k,j,i}(y_i) \\
\leq{} & z_{i,2}z_{i,3} - \sum_{j=1}^{2}c_{i,i}z_{i,i}^2 + z_{i,2}\frac{\partial\alpha_{i,1}}{\partial\hat{\theta}_i}\left(\Gamma_i\tau_{i,2} - \Gamma_i l_{\theta i}(\hat{\theta}_i - \theta_{i0}) - \dot{\hat{\theta}}_i\right) \\
& -\tilde{\theta}_i^T\left(\Gamma_i^{-1}\dot{\hat{\theta}}_i - \tau_{i,2}\right) + \sum_{j=1}^{2}(3 - j)\sum_{k=1}^{N} r_{i,j,k}\bar{h}_{i,j,k}(y_k) \\
& -z_{i,1}sg_i(z_{i,1})\sum_{j=1}^{n_i}(n_i - j + 1)\sum_{k=1}^{N} r_{i,j,k}\bar{h}_{k,j,i}(y_i) \qquad (6.23)
\end{aligned}
$$

Step i, ($i = 3, \ldots, n_i$). Repeating the procedure in a recursive manner, the actual control input u_i appears and is at our disposal in the last step n_i.

The decentralized adaptive backstepping controllers are summarized in Table 6.1. In the control scheme (6.24)–(6.33), $c_{i,j}$, $l_{\theta i}$, σ_i, λ_i, and μ_i are positive constants, $\mu_i \geq u_{min,i}$, Γ_i is a positive definite matrix, and $\hat{\theta}_i$ is an estimate of θ_i, and θ_{i0} is a constant, $i = 1, \ldots, N$, $j = 1, \ldots, n_i$.

6.4 Stability Analysis

The main results are formally stated in the following theorem.

Theorem 6.1

Consider the interconnected systems (6.2) with input quantization and nonlinear interconnections under Assumptions 6.1–6.2, the decentralized adaptive backstepping controller (6.26) with virtual control laws (6.27)–(6.29) and the parameter estimator with updating law (6.33), the following results can be guaranteed.

1. *All the closed-loop signals are globally uniformly bounded.*

2. *The tracking error signals $e(t) = [e_1, e_2, \ldots e_N]^T$, where $e_i = y_i - y_{ri}$ for $i = 1, 2, \ldots, N$ will converge to a compact set.*

Proof 6.1 In the last step n_i of subsystem i, similar to the derivation in Chapter 5, the following inequality is derived by multiplying $|z_{i,n_i}|$ on both sides of (6.4) and using (6.26) and (6.27),

$$
\begin{aligned}
z_{i,n_i} q_i(u_i) \leq\ & z_{i,n_i} u_i + \delta_i |z_{i,n_i} u_i| + u_{min,i} |z_{i,n_i}| \\
\leq\ & -z_{i,n_i} u_{n_i} \tanh(z_{i,n_i} u_{n_i}/\lambda_i) + \delta_i |z_{i,n_i} u_{n_i} \tanh(z_{i,n_i} u_{n_i}/\lambda)| \\
& + u_{min,i} |z_{i,n_i}| \\
\leq\ & -(1 - \delta_i) z_{i,n_i} u_{i,n} \tanh(z_{i,n_i} u_{n_i}/\lambda_i) + u_{min,i} |z_{i,n_i}| \\
\leq\ & -(1 - \delta_i) |z_{i,n_i} u_{n_i}| + u_{min,i} |z_{i,n_i}| + \epsilon_{i1} \\
\leq\ & -(1 - \delta_i)(z_{i,n_i} u_{n_i}) + \mu_i |z_{i,n_i}| + \epsilon_{i1} \\
\leq\ & z_{i,n_i} \alpha_{i,n_i} - \mu_i z_{i,n_i} \tanh(\mu_i z_{i,n_i}/\lambda_i) + \mu_i |z_{i,n}| + \epsilon_{i1} \\
\leq\ & z_{i,n_i} \alpha_{i,n} + \epsilon_{i2},
\end{aligned} \tag{6.34}
$$

where μ_i is a positive constant such that $\mu_i \geq u_{min,i}$, $\epsilon_{i,1} = 0.2785\lambda_i(1 - \delta_i)$, and $\epsilon_{i2} = 0.2785\lambda_i(2 - \delta_i)$.

For subsystem i, we choose the local Lyapunov function as

$$
U_{i,n_i} = \sum_{j=1}^{n_i} \frac{1}{2} z_{i,j}^2 + \frac{1}{2} \tilde{\theta}_i^T \Gamma_i^{-1} \tilde{\theta}_i. \tag{6.35}
$$

Table 6.1 Decentralized Adaptive Backstepping Control Scheme

Coordinate transformation:

$$z_{i,1} = y_i - y_{ri} \tag{6.24}$$

$$z_{i,j} = x_{i,j} - \alpha_{i,j-1}, \ j = 2, 3, \ldots, n_i \tag{6.25}$$

Control laws:

$$u_i = -\tanh(z_{i,n_i} u_{n_i}/\lambda_i) u_{n_i} \tag{6.26}$$

$$u_{n_i} = \frac{1}{1 - \delta_i} \left(-\alpha_{i,n_i} + \mu_i \tanh(\mu_i z_{i,n_i}/\lambda_i) \right) \tag{6.27}$$

$$\alpha_{i,1} = -c_{i,1} z_{i,1} - \frac{1}{4} z_{i,1} - \psi_{i,1} - \omega_{i,1}^T \hat{\theta}_i + \dot{y}_{ri}$$

$$- sg_i(z_{i,1}) \sum_{j=1}^{n_i} (n_i - j + 1) \sum_{k=1}^{N} r_{i,j,k} \bar{h}_{k,j,i}(y_i) \tag{6.28}$$

$$\alpha_{i,j} = -c_{i,j} z_{i,j} - \frac{1}{4} z_{i,j} - \frac{1}{4} \sum_{k=1}^{j-1} \left(\frac{\partial \alpha_{i,j-1}}{\partial x_{i,k}} \right)^2 z_{i,j} - z_{i,j-1}$$

$$- \psi_{i,j} - \omega_{i,j}^T \hat{\theta}_i + \sum_{k=1}^{j-1} \frac{\partial \alpha_{i,j-1}}{\partial x_{i,k}} (x_{i,k+1} + \psi_{i,k})$$

$$+ \frac{\partial \alpha_{i,j-1}}{\partial \hat{\theta}_i} \left(\Gamma_i \tau_{i,j} - \Gamma_i l_{\theta i}(\hat{\theta}_i - \theta_{i0}) \right)$$

$$+ \sum_{k=2}^{j-1} \frac{\partial \alpha_{i,k-1}}{\partial \hat{\theta}_i} \Gamma_i \omega_{i,j} z_k + y_{ri}^{(j)} \tag{6.29}$$

with

$$\tau_{i,j} = \tau_{i,j-1} + \omega_{i,j} z_{i,j} \tag{6.30}$$

$$\omega_{i,j} = \phi_{i,j} - \sum_{k=1}^{j} \frac{\partial \alpha_{i,j-1}}{\partial x_{i,k}} \phi_{i,k}, \ j = 1, \ldots, n_i \tag{6.31}$$

$$sg_i(z_{i,1}) = \begin{cases} \frac{1}{z_{i,1}} \frac{z_{i,1}}{(z_{i,1}^2 - \sigma_i^2)^{n_i} + \sigma_i} & |z_{i,1}| \geq \sigma_i \\ & |z_{i,1}| < \sigma_i \end{cases} \tag{6.32}$$

Parameter update law:

$$\dot{\hat{\theta}}_i = \Gamma_i \tau_{i,n_i} - \Gamma_i l_{\theta i}(\hat{\theta}_i - \theta_{i0}) \tag{6.33}$$

The derivative of U_{i,n_i} is given by

$$
\begin{aligned}
\dot{U}_{i,n_i} \leq &-\sum_{j=1}^{n_i} c_{i,j} z_{i,j}^2 + \epsilon_{i,2} - \tilde{\theta}_i^T \left(\Gamma_i^{-1} \dot{\hat{\theta}}_i - \tau_{n_i} \right) \\
&+ \left(\sum_{j=2}^{n_i} z_{i,j} \frac{\partial \alpha_{i,j-1}}{\partial \hat{\theta}_i} \right) \left(\Gamma_i \tau_{i,n} - \Gamma_i l_{\theta i} (\hat{\theta}_i - \theta_{i0}) - \dot{\hat{\theta}}_i \right) \\
&+ \sum_{j=1}^{n_i} z_{i,j} \left(h_{i,j} + \sum_{k=1}^{j-1} \frac{\partial \alpha_{i,j-1}}{\partial x_{i,k}} h_{i,k} \right) \\
&- \sum_{j=1}^{n_i} \left(\frac{1}{4} z_{i,j}^2 + \frac{1}{4} \sum_{k=1}^{j-1} \left(\frac{\partial \alpha_{i,j-1}}{\partial x_{i,k}} \right)^2 z_{i,j}^2 \right) \\
&- sg_i(z_{i,1}) z_{i,1} \sum_{j=1}^{n_i} (n_i - j + 1) \sum_{k=1}^{N} r_{i,j,k} \bar{h}_{k,j,i}(y_i),
\end{aligned} \tag{6.36}
$$

Using the following property

$$
l_{\theta i} \tilde{\theta}_i (\hat{\theta}_i - \theta_{i0}) \leq -\frac{1}{2} l_{\theta i} \parallel \tilde{\theta}_i \parallel^2 + \frac{1}{2} l_{\theta i} \parallel (\theta_i - \theta_{i0}) \parallel^2, \tag{6.37}
$$

and applying the parameter updating law (6.33), we have

$$
\begin{aligned}
&-\tilde{\theta}_i^T \left(\Gamma_i^{-1} \dot{\hat{\theta}}_i - \tau_{n_i} \right) + \left(\sum_{j=2}^{n_i} z_{i,j} \frac{\partial \alpha_{i,j-1}}{\partial \hat{\theta}_i} \right) \left(\Gamma_i \tau_{i,n} - \Gamma_i l_{\theta i} (\hat{\theta}_i - \theta_{i0}) - \dot{\hat{\theta}}_i \right) \\
&\leq -\frac{1}{2} l_{\theta i} \parallel \tilde{\theta}_i \parallel^2 + \frac{1}{2} l_{\theta i} \parallel \theta_i - \theta_{i0} \parallel^2
\end{aligned} \tag{6.38}
$$

Then the derivative of U_{i,n_i} follows

$$
\begin{aligned}
\dot{U}_{i,n_i} \leq &-\sum_{j=1}^{n_i} c_{i,j} z_{i,j}^2 - \frac{1}{2} l_{\theta i} \parallel \tilde{\theta}_i \parallel^2 + \frac{1}{2} l_{\theta i} \parallel \theta_i - \theta_{i0} \parallel^2 \\
&+ \epsilon_{i,2} + \sum_{j=1}^{n_i} z_{i,j} \left(h_{i,j} + \sum_{k=1}^{j-1} \frac{\partial \alpha_{i,j-1}}{\partial x_{i,k}} h_{i,k} \right) \\
&- \sum_{j=1}^{n_i} \left(\frac{1}{4} z_{i,j}^2 + \frac{1}{4} \sum_{k=1}^{j-1} \left(\frac{\partial \alpha_{i,j-1}}{\partial x_{i,k}} \right)^2 z_{i,j}^2 \right) \\
&- sg_i(z_{i,1}) z_{i,1} \sum_{j=1}^{n_i} (n_i - j + 1) \sum_{k=1}^{N} r_{i,j,k} \bar{h}_{k,j,i}(y_i)
\end{aligned} \tag{6.39}
$$

Let $F_i = \min\{2c_{i,1}, 2c_{i,2}, \ldots, 2c_{i,n_i}, l_{\theta i} \lambda_{min}(\Gamma_i)\}$, it gives

$$
\begin{aligned}
U_{i,n_i} \leq & -F_i U_{i,n_i} + \frac{1}{2} l_{\theta i} \parallel \theta_i - \theta_{i0} \parallel^2 + \epsilon_{i,2} \\
& - sg_i(z_{i,1}) z_{i,1} \sum_{j=1}^{n_i} (n_i - j + 1) \sum_{k=1}^{N} r_{k,j,i} \bar{h}_{k,j,i}(y_i) \\
& + \sum_{j=1}^{n_i} (n_i - j + 1) \sum_{k=1}^{N} r_{i,j,k} \bar{h}_{i,j,k}(y_k),
\end{aligned}
\tag{6.40}
$$

where Young's inequality and (6.3) are applied, by noting that

$$
\begin{aligned}
& \sum_{j=1}^{n_i} z_{i,j} \left(h_{i,j} + \sum_{k=1}^{j-1} \frac{\partial \alpha_{i,j-1}}{\partial x_{i,k}} h_{i,k} \right) - \sum_{j=1}^{n_i} \left(\frac{1}{4} + \frac{1}{4} \sum_{k=1}^{j-1} \left(\frac{\partial \alpha_{i,j-1}}{\partial x_{i,k}} \right)^2 \right) z_{i,j}^2 \\
& \leq \sum_{j=1}^{n_i} (n_i - j + 1)(h_{i,j}(y_1, \ldots, y_N))^2 \\
& \leq \sum_{j=1}^{n_i} (n_i - j + 1) \sum_{k=1}^{N} r_{i,j,k} \bar{h}_{i,j,k}(y_k).
\end{aligned}
\tag{6.41}
$$

Choose an overall Lyapunov function for the entire group of subsystems as

$$
U = \sum_{i=1}^{N} U_{i,n_i}.
\tag{6.42}
$$

From (6.26)–(6.40), we obtain

$$
\dot{U} \leq - \sum_{i=1}^{N} F_i U_{i,n_i} + \sum_{i=1}^{N} \left(\frac{1}{2} l_{\theta i} \parallel \theta_i - \theta_{i0} \parallel^2 + \epsilon_{i,2} \right) + H
\tag{6.43}
$$

where

$$
\begin{aligned}
H = & -\sum_{i=1}^{N} \sum_{j=1}^{n_i} \sum_{k=1}^{N} sg_i(z_{i,1}) z_{i,1} (n_i - j + 1) r_{k,j,i} \bar{h}_{k,j,i}(y_i) \\
& + \sum_{i=1}^{N} \sum_{j=1}^{n_i} \sum_{k=1}^{N} (n_i - j + 1) r_{i,j,k} |\bar{h}_{i,j,k}(y_k)|.
\end{aligned}
\tag{6.44}
$$

From the definition of sg_i in (6.32), it is clear that $H = 0$ for $|z_{i,1}| \geq \sigma_i$. For $|z_{i,1}| < \sigma_i$,

$$
H \leq \sum_{i=1}^{N} \sum_{j=1}^{n_i} \sum_{k=1}^{N} (n_i - j + 1) r_{i,j,k} \bar{h}_{i,j,k}(z_{i,1} + y_{ri}).
\tag{6.45}
$$

Clearly H has an upper bound $\bar{H} \geq 0$ from the boundedness of y_{ri} and $|z_{i,1}| < \sigma_i$. It follows that

$$\dot{U} \leq -FU + M, \tag{6.46}$$

where $F = \min\{F_i\}, i = 1, \ldots, N$, $M = \sum_{i=1}^{N} \left(\frac{1}{2}l_{\theta i} \parallel \theta_i - \theta_{i0} \parallel^2 + \epsilon_{i,2} \right) + \bar{H}$. By direct integration of the differential inequality (6.46), we have

$$U \leq U(0)e^{-Ft} + \frac{M}{F}(1 - e^{-Ft}), \tag{6.47}$$

where $U(0) = \sum_{i=1}^{N} \sum_{j=1}^{n_i} \frac{1}{2}z_{i,j}^2(0) + \frac{1}{2}\tilde{\theta}_i^T(0)\Gamma_i^{-1}\tilde{\theta}_i(0)$. It shows that U is uniformly bounded, yielding that $x_{i,j}, \hat{\theta}_i, \alpha_{i,j}, u_i$ for $i = 1, \ldots, N$ and $j = 1, \ldots, n_i$ are all bounded. From the definitions of $e(t)$, U and (6.47), we obtain that

$$\parallel e(t) \parallel^2 \leq \max\left\{ 2U(0), \frac{2M}{F} \right\}. \tag{6.48}$$

It implies that the tracking errors will converge to a compact set.

Remark 6.2   The main difficulties are to handle the unknown interactions in all the differential equations of the subsystems. At each control step j, in addition to compensate for the interaction $h_{i,j}$, we also need to compensate for the interactions $h_{i,k}, k = 1, .., j - 1$ from the previous differential equation of $x_{i,k}$ by using backstepping technique. In order to handle these effects, a new compensation scheme is constructed by introducing a well-defined smooth function in (6.32) and new terms in the controller (6.28) and (6.29). Compared with the adaptive controller designed for single-loop nonlinear systems with input quantization, the new term $-sg_i(z_{i,1})\sum_{j=1}^{n_i}(n_i - j + 1)\sum_{k=1}^{N} r_{i,j,k}\bar{h}_{k,j,i}(y_i)$ is introduced in the local control law $\alpha_{i,1}$ in (6.28) to compensate for the effects of interactions $h_{i,j}$ among other subsystems $j \neq i$. The other new terms $-\frac{1}{4}z_{i,j}$ and $-\frac{1}{4}\sum_{k=1}^{j-1}\left(\frac{\partial \alpha_{i,j-1}}{\partial x_{i,k}}\right)^2 z_{i,j}$ in the local virtual control laws (6.28) and (6.29) are used to compensate for the effects from the unmodeled part $h_{i,j}$ of its own subsystem $j = i$. Note that a well-defined function $sg_i(z_{i,1})$ in (6.32) is continuous and n_i th-order differentiable.

Remark 6.3   Note that the interaction $h_{i,j}$ at each equations of the interconnected systems (6.2) is transformed to a term $\sum_{j=1}^{n_i}(n_i - j + 1)\sum_{k=1}^{N} r_{i,j,k}\bar{h}_{i,j,k}(y_k)$ in (6.40). The key steps are (6.41) and (6.44) in the stability analysis, which results in the cancellation of the interaction effects from other subsystems.

Remark 6.4   In [101], the original system was transformed to a system including only the output signal and the signals from filters. Only interactions $h_{i,1}$ exist in the equation for the output in the final control systems and the rest equations related to the filter signals do not involve interactions. So the designed controller only need to compensate for the effects from the interaction $h_{i,1}$. Compared to [101], the class of interconnected systems given in (6.2) of this paper is more general, as interactions exist in all the differential equations of subsystems. From the proposed scheme, it can be noted that at step j of the backstepping design, we need not only to compensate for the effects of interaction $h_{i,j}$ from the jth equation, but also the effects of interactions $h_{i,k}$ $(k = 1, \ldots, j-1)$ from the previous $j-1$ equations. Thus for such systems, it is more challenging to design appropriate controllers to compensate for the effects of unknown interactions.

6.5 An Illustrative Example

An interconnected system is considered with two subsystems and hysteresis-type quantized inputs as follows:

$$\begin{aligned}
\ddot{y}_1 + \theta_1\phi_1 + h_1 &= q_1(u_1) \\
\ddot{y}_2 + \theta_2\phi_2 + h_2 &= q_2(u_2),
\end{aligned} \tag{6.49}$$

where $\phi_1 = y_1^2$, $\phi_2 = y_2 + y_2^2$, the interconnection terms $h_1 \le y_2 + y_1$, $h_2 \le y_1^2 + y_2$, the parameters θ_1 and θ_2 are unknown. $q_1(u_1)$ and $q_2(u_2)$ represent a hysteresis-logarithmic quantizer modeled in (3.4), where δ_i and $u_{min,i}$ for $i = 1, 2, \ldots$ are quantization parameters. In the simulation, the tracking trajectories are $y_{r1} = sin(t)$ and $y_{r2} = 1 - cos(t)$, the quantization parameters are chosen as $\delta_i = 0.2$ and $u_{min,i} = 0.1$, the initial conditions are $[x_1(0), \dot{x}_1(0)] = [0.5, 0]^T$, $[x_2(0), \dot{x}_2(0)] = [0.5, 0]^T$ and $\hat{\theta}_i(0) = 0.8$. The design parameters are chosen as $c_{i1} = 6, c_{i2} = 4$, $\Gamma_i = 1, l_{\theta i} = 0.01, \theta_{i0} = 0.01, \lambda_i = 0.2, \mu_i = 0.1$. The responses of all subsystem outputs $y_i(t)$ and control inputs u_i are shown in Figures 6.1–6.2. Clearly, all these signals are bounded which is in accordance with the theoretical findings in Theorem 6.1.

6.6 Notes

In this chapter, adaptive tracking of a class of uncertain interconnected systems with input quantization is considered. A totally decentralized adaptive backstepping control scheme is developed together with a new compensation method constructed for the unknown nonlinear interactions. By introducing a hyperbolic tangent function and a new transformation of the final control signal, the effects from the input quantization are effectively compensated. It is established that the proposed decentralized controllers can ensure global stability of the overall system and the transient performance of the tracking errors can be improved by appropriately tuning design parameters.

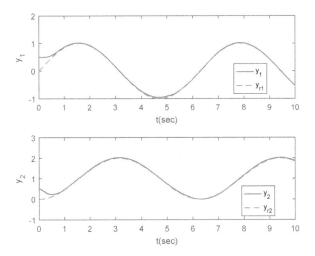

Figure 6.1 Outputs y_1 and y_2 of interconnected system with hysteresis-logarithmic quantizer

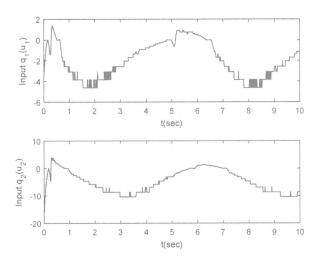

Figure 6.2 Inputs $q_1(u_1)$ and $q_2(u_2)$ of interconnected system with hysteresis-logarithmic quantizer

Acknowledgment

Reprinted from Copyright (2021), with permission from Elsevier. Jing Zhou, Changyun Wen and Wei Wang, "Adaptive control of uncertain nonlinear systems with quantized input signal", *Automatica*, vol. 95, pp. 152–162, 2018.

Chapter 7

Output Feedback Control for Uncertain Nonlinear Systems with Input Quantization

Chapters 4–6 have presented several adaptive control schemes for uncertain nonlinear systems with input quantization. Yet, all these results are based on state-feedback control, i.e. all the states of the considered systems are measurable and thus available for controller design. However, for many practical systems, only output signals can be measured. This chapter will investigate adaptive output-feedback tracking control of a class of uncertain nonlinear systems with quantized input signal. Two types of quantizers are considered here, i.e. quantizers with bounded property and quantizers with sector-bounded property as given in Chapter 3. For these two kinds of quantizers, different control schemes are presented. Compared with existing results in adaptive control, the proposed schemes provide a way to relax certain restrictive conditions, in addition to solving the problem of adaptive output-feedback control with input quantization. It is shown that the designed adaptive controller ensures global boundedness of all the signals in the closed-loop system and enables the tracking error to exponentially converge towards a compact set which is adjustable.

DOI: 10.1201/9781003176626-7

7.1 Introduction

In Chapters 4-6, several adaptive state-feedback control schemes are presented for uncertain nonlinear system with input quantization, where all states are assumed to be avaiable for controller design. However, in many control engineering systems, only output signals can be measured while the other state signals cannot. For this class of systems, output-feedback control is needed. Therefore, it is also important to investigate output-feedback control of systems with input quantization.

Compared with quantized control results through state-feedback control, output-feedback control results are still very limited. The main difficulties lies in the following two aspects:

■ **Challenge 1.** How to design suitable state observers when input signal is quantized?

■ **Challenge 2.** How to handle the quantization errors?

In this chapter, we present adaptive output-feedback control schemes for a class of uncertain nonlinear systems with input quantization. The considered system contains unknown parameters and only the output signal is measurable. Two different schemes addressing different quantizers are proposed. Specifically, the first control scheme deals with quantizers with bounded quantization errors, such as the logarithmic-uniform quantizer (3.5) in Chapter 3. The second scheme deals with quantizers with unbounded quantization errors, and thus all the sector-bounded quantizers introduced in Chapter 3 can be included. For the first challenge, proper state observers are designed using the input signal u in the first scheme, and the quantized $q(u)$ in the second scheme. Based on the observers, different control laws are designed to compensate for the quantization errors such that Challenge 2 is also overcome.

7.2 Problem Formulation

The following class of uncertain nonlinear systems is considered.

$$\dot{x}_1 = x_2 + \phi_1^T(y)\theta + \varphi_1(y)$$

$$\vdots \quad \vdots$$

$$\dot{x}_{\zeta-1} = x_\zeta + \phi_{\zeta-1}^T(y)\theta + \varphi_{\zeta-1}(y)$$

$$\dot{x}_\zeta = x_{\zeta+1} + \phi_\zeta^T(y)\theta + \varphi_\zeta(y) + b_m q_s(u)$$

$$\vdots \quad \vdots$$

$$\dot{x}_{n-1} = x_n + \phi_{n-1}^T(y)\theta + \varphi_{n-1}(y) + b_1 q_s(u)$$

$$\dot{x}_n = \phi_n^T(y)\theta + \varphi_n(y) + b_0 q_s(u)$$

$$y = x_1(t) \tag{7.1}$$

where $x_1, \ldots, x_n, y$ and u are system states, output and control input signal to be designed. $\theta \in \mathbb{R}^r$ and $b_m, \ldots, b_0$ are unknown constants. $\phi_i(y) \in \mathbb{R}^r$ and $\varphi_i(y) \in \mathbb{R}^1$, i=1,...,n are known smooth nonlinear functions. $q_s(u)$ is the quantized input signal and take quantized values.

The objective of this paper is to propose a control design scheme which can make the output $y = x_1(t)$ track a reference signal $r(t)$ with the input quantized by a quantizer. We assume that the existence and uniqueness of a solution forward in time are satisfied for such a class of nonlinear systems.

For the development of the control laws, the following assumptions are made.

Assumption 7.1 *The sign of b_m is known.*

Assumption 7.2 *The relative degree $\zeta = n - m$ is known and the system is minimum phase, i.e. the polynomial $B(s) = b_m s^m + \cdots + b_1 s + b_0$ is Hurwitz.*

Assumption 7.3 *The reference signal y_r and its first ζth order derivatives are piecewise continuous, known and bounded.*

7.3 Controller Design with Bounded Quantizers

In this section, we present the controller design method for bounded quantizers, i.e. quantizers which always yield bounded quantization errors. Typical examples include uniform quantizer (3.1) hysteresis-uniform quantizer (3.3), and logarithmic-uniform quantizer (3.5) in Chapter 3.

7.3.1 State Estimation Filters

In order to design the desired adaptive output-feedback control law, we rewrite system (7.1) in the following form

$$\dot{x} = Ax + \Phi(y)\theta + \Psi(y) + \begin{pmatrix} 0 \\ b \end{pmatrix} q(u) \tag{7.2}$$

where

$$A = \begin{pmatrix} 0 & 1 & 0 & \cdots & 0 \\ 0 & 0 & 1 & \cdots & 0 \\ \vdots & \vdots & \vdots & \ddots & \vdots \\ 0 & 0 & 0 & \cdots & 1 \\ 0 & 0 & 0 & \cdots & 0 \end{pmatrix} \qquad \Psi(y) = \begin{pmatrix} \varphi_1(y) \\ \vdots \\ \varphi_n(y) \end{pmatrix} \tag{7.3}$$

$$\Phi(y) = \begin{pmatrix} \phi_1^T(y) \\ \vdots \\ \phi_n^T(y) \end{pmatrix} \qquad b = \begin{pmatrix} b_m \\ \vdots \\ b_0 \end{pmatrix} \tag{7.4}$$

Since x is unavailable and only the output y is measurable, we need to design filters to estimate x. With the input signal u quantized by a bounded quantizer, the filters are designed as follows:

$$\dot{\xi} = A_0\xi + ky + \Psi(y) \tag{7.5}$$

$$\dot{\Xi}^T = A_0\Xi^T + \Phi(y) \tag{7.6}$$

$$\dot{\lambda} = A_0\lambda + e_n u \tag{7.7}$$

$$\nu_i = A_0^i\lambda, \quad i = 0, 1, \dots m \tag{7.8}$$

where $e_i(i = 1, \dots, n)$ is a row vector with the ith entry being 1 and the others being 0, $k = [k_1, \dots, k_n]^T$ such that all eigenvalues of $A_0 = A - ke_1^T$ are at some desired stable locations.

The state estimations are given by

$$\hat{x}(t) = \xi + \Xi^T\theta + \sum_{i=0}^{m} b_i\nu_i \tag{7.9}$$

Then we get the derivative of $\hat{x}$ as

$$\dot{\hat{x}}(t) = A_0\hat{x} + ky + \Phi(y)\theta + \Psi(y) + \begin{pmatrix} 0 \\ b \end{pmatrix}u \tag{7.10}$$

The state estimation error is defined as

$$\epsilon = x(t) - \hat{x}(t) \tag{7.11}$$

Remark 7.1 Note that all states of the filters in (7.5)–(7.8) are available for feedback. However the estimated states given in (7.11) are still unknown due to unknown parameters and thus cannot be utilized in controller design. Instead, they will be used for analysis of the resulting closed-loop system. The following Lemma shows a property of the proposed state observers.

Lemma 7.1
For system (7.1), the proposed filters (7.5)–(7.8) guarantee that the state estimation error is always bounded for all $t > 0$, regardless of the control signal.

Proof: From (7.1) and (7.5)–(7.11), the state estimation error satisfies

$$\begin{aligned} \dot{\epsilon} &= A_0\epsilon + \begin{pmatrix} 0 \\ b \end{pmatrix}(q(u) - u) \\ &= A_0\epsilon + B\Delta_q \end{aligned} \tag{7.12}$$

where $B = \begin{pmatrix} 0 \\ b \end{pmatrix}$ and $\Delta_q = q(u) - u$.

Since A_0 is Hurwitz, there must exist a positive definite matrix P satisfying $A_0^T P + P A_0 \leq -I$. By considering the Lyapunov function $V_\epsilon = \frac{1}{2}\epsilon^T P \epsilon$, we can obtain

$$
\begin{aligned}
\dot{V}_\epsilon &= \frac{1}{2}\epsilon^T(A_0^T P + P A_0)\epsilon + \epsilon^T P B \Delta_q \\
&\leq (-\frac{1}{2} + \frac{||P||}{4\alpha})||\epsilon||^2 + \alpha||P|| \, ||B||^2 \Delta_q^2 \\
&\leq -\frac{\beta}{\lambda_{max}(P)}V_\epsilon + \alpha||P|| \, ||B||^2 \Delta_q^2
\end{aligned}
\tag{7.13}
$$

where $\alpha > 0$ satisfying $\beta = \frac{1}{2} - \frac{||P||}{4\alpha} > 0$. Let $\lambda_{max}(P)$ and $\lambda_{min}(P)$ denote the the biggest and smallest eigenvalues of the matrix P, respectively. Then we have

$$
\begin{aligned}
\lambda_{min}(P)||\epsilon||^2 &\leq V_\epsilon \leq e^{-\frac{\beta}{\lambda_{max}(P)}t}V_\epsilon(0) \\
&+ \frac{\alpha||P|| \, ||B||^2 \lambda_{max}(P)\Delta_q^2}{\beta}(1 - e^{-\frac{\beta}{\lambda_{max}(P)}t})
\end{aligned}
\tag{7.14}
$$

So the size of the state estimation error $||\epsilon||$ converges exponentially towards a bounded compact set $\Omega_\epsilon = \{\epsilon| \, ||\epsilon||^2 \leq \frac{\alpha||P|| \, ||B||^2 \lambda_{max}(P)\Delta_q^2}{\beta \lambda_{min}(P)}\}$ at a rate of $\frac{\beta}{\lambda_{max}(P)}$. Note that the size of Ω_ϵ is determined by the quantization error Δ_q. Since Δ_q is always bounded for bounded quantizers, the state estimation error is always bounded, independent of the control signal u. ∎

7.3.2 Design of Adaptive Controller

In this section, we will design an adaptive controller based on backstepping technique with tuning functions, which involves ζ recursive steps.

Let ϵ_2, $\upsilon_{i,2}$, ξ_2, and Ξ_2 denote the second entries of ϵ, υ_i, ξ, and Ξ, respectively, and $\upsilon_{m,i}$ denote the ith enrty of υ_m. Then, we have

$$
\begin{aligned}
\dot{y} &= b_m \upsilon_{m,2} + \xi_2 + \varphi_1(y) + \bar{w}^T\Theta + \epsilon_2 & (7.15)\\
\dot{\upsilon}_{m,i} &= \upsilon_{m,i+1} - k_i\upsilon_{m,1}, \quad i = 2,3,\ldots,\zeta-1 & (7.16)\\
\dot{\upsilon}_{m,\zeta} &= \upsilon_{m,\zeta+1} - k_\zeta\upsilon_{m,1} + u & (7.17)
\end{aligned}
$$

where

$$
\begin{aligned}
\Theta &= [b_m,\ldots,b_0,\theta^T]^T & (7.18)\\
w &= [\upsilon_{m,2},\upsilon_{m-1,2},\ldots,\upsilon_{0,2},\Xi_2 + \phi_1^T]^T & (7.19)\\
\bar{w} &= [0,\upsilon_{m-1,2},\ldots,\upsilon_{0,2},\Xi_2 + \phi_1^T]^T & (7.20)
\end{aligned}
$$

Then we take the change of coordinates

$$
\begin{aligned}
z_1 &= y - y_r & (7.21)\\
z_i &= \upsilon_{m,i} - \alpha_{i-1} - \hat{\rho}y_r^{(i-1)} \quad i = 2,\ldots,\zeta & (7.22)
\end{aligned}
$$

where $\hat{\rho}$ is the estimation of $\rho = \frac{1}{b_m}$, and α_{i-1} is the virtual control at step $i-1$. The ζ design steps are summarized as follows by following the recursive backstepping procedure.

Step 1: Select the virtual control law α_1 as

$$\alpha_1 = \hat{\rho}\bar{\alpha}_1 \tag{7.23}$$

$$\bar{\alpha}_1 = -c_1 z_1 - d_1 z_1 - \xi_2 - \varphi_1(y) - \bar{w}\hat{\Theta} \tag{7.24}$$

where c_1 and d_1 are positive design parameters. Then we have

$$
\begin{aligned}
\dot{z}_1 = {} & -(c_1 + d_1)z_1 + \epsilon_2 + (w - \hat{\rho}(\dot{y}_r + \bar{\alpha}_1)e_1)^T \tilde{\Theta} \\
& -b_m(\dot{y}_r + \bar{\alpha}_1)\tilde{\rho} + \hat{b}_m z_2
\end{aligned} \tag{7.25}
$$

Choose the Lyapunov function V_1 as

$$V_1 = \frac{1}{2}z_1^2 + \frac{1}{2}\tilde{\Theta}^T \Gamma^{-1}\tilde{\Theta} + \frac{|b_m|}{2\gamma}\tilde{\rho}^2 \tag{7.26}$$

The updating law of $\hat{\rho}$ is chosen as

$$\dot{\hat{\rho}} = -\gamma sign(b_m)(\dot{y}_r + \bar{\alpha}_1)z_1 - \sigma_1 \hat{\rho} \tag{7.27}$$

where γ and σ_1 are positive constants and Γ^{-1} is a positive definite matrix. Define

$$\tau_1 = (w - \hat{\rho}(\dot{y}_r + \bar{\alpha}_1)e_1)z_1 \tag{7.28}$$

Then we get

$$
\begin{aligned}
\dot{V}_1 \leq {} & -c_1 z_1^2 + \hat{b}_m z_1 z_2 + \tilde{\Theta}^T(\tau_1 - \Gamma^{-1}\dot{\hat{\Theta}}) \\
& +\frac{|b_m|}{\gamma}\sigma_1\tilde{\rho}\hat{\rho} + \frac{\epsilon^T\epsilon}{4d_1}
\end{aligned} \tag{7.29}
$$

Step 2: We choose the second virtual control law α_2 and the tuning function as

$$\alpha_2 = -\hat{b}_m z_1 - (c_2 + d_2(\frac{\partial\alpha_1}{\partial y})^2)z_2 + \beta_2 + \frac{\partial\alpha_1}{\partial\hat{\Theta}}\Gamma(\tau_2 - \sigma_2\hat{\Theta}) \tag{7.30}$$

$$\tau_2 = \tau_1 - \frac{\partial\alpha_1}{\partial y}wz_2 \tag{7.31}$$

where c_2, σ_2, and d_2 are positive constants and

$$
\begin{aligned}
\beta_2 = {} & \frac{\partial\alpha_1}{\partial y}(\xi_2 + \psi_1 + w^T\hat{\Theta}) + k_2 v_{m,1} + \frac{\partial\alpha_1}{\partial y_r}\dot{y}_r \\
& + (\dot{y}_r + \frac{\partial\bar{\alpha}_1}{\hat{\rho}})\dot{\hat{\rho}} + \sum_{j=1}^{m+i-1}\frac{\partial\alpha_1}{\lambda_j}(-k_j\lambda_1 + \lambda_{j+1}) \\
& + \frac{\partial\alpha_1}{\partial\xi}(A_0\xi + ky + \Psi(y)) + \frac{\partial\alpha_1}{\partial\Xi^T}(A_0\Xi^T + \Phi(y))
\end{aligned} \tag{7.32}
$$

Choose Lyapunov function as

$$V_2 = V_1 + \frac{1}{2}z_2^2 \tag{7.33}$$

Then we obtain

$$
\begin{aligned}
\dot{V}_2 \leq & -\sum_{i=1}^{2} c_i z_i^2 + z_2 z_3 + \tilde{\Theta}^T (\tau_2 - \Gamma^{-1}\dot{\hat{\Theta}}) + \frac{|b_m|}{\gamma}\sigma_1 \tilde{\rho}\hat{\rho} \\
& + \frac{\partial \alpha_1}{\partial \hat{\Theta}}(\Gamma\tau_2 - \Gamma\sigma_2\hat{\Theta} - \dot{\hat{\Theta}}) + (\frac{1}{4d_1} + \frac{1}{4d_2})\epsilon^T \epsilon
\end{aligned}
\tag{7.34}
$$

Step i ($i = 3, \ldots, \zeta$): Choose the virtual control law and the tuning function as

$$
\begin{aligned}
\alpha_i = & -z_{i-1} - [c_i + d_i(\frac{\partial \alpha_{i-1}}{\partial y})^2]z_i + \beta_i + \frac{\partial \alpha_{i-1}}{\partial \hat{\Theta}}\Gamma(\tau_i - \sigma_2\hat{\Theta}) \\
& -(\Sigma_{k=2}^{i-1} z_k \frac{\partial \alpha_{k-1}}{\hat{\Theta}})\Gamma\frac{\partial \alpha_{i-1}}{\partial y}\omega, \quad i = 3, \ldots, \zeta
\end{aligned}
\tag{7.35}
$$

where c_i and d_i are positive constants and

$$\tau_i = \tau_{i-1} - \frac{\partial \alpha_{i-1}}{\partial y}\omega z_i \tag{7.36}$$

$$
\begin{aligned}
\beta_i = & \frac{\partial \alpha_{i-1}}{\partial y}(\xi_2 + \psi_1 + \omega^T\hat{\Theta}) + k_i v_{m,1} + \sum_{j=1}^{i-1} \frac{\partial \alpha_{i-1}}{\partial y_r^{(j-1)}}y_r^{(j)} \\
& + (y_r^{(i-1)} + \frac{\partial \alpha_{i-1}}{\hat{\rho}})\dot{\hat{\rho}} + \sum_{j=1}^{m+i-1} \frac{\partial \alpha_{i-1}}{\lambda_j}(-k_j\lambda_1 + \lambda_{j+1}) \\
& + \frac{\partial \alpha_{i-1}}{\partial \xi}(A_0\xi + ky + \Psi(y)) + \frac{\partial \alpha_{i-1}}{\partial \Xi^T}(A_0\Xi^T + \Phi(y))
\end{aligned}
\tag{7.37}
$$

In the last design step, the adaptive controller and the parameter updating law are finally obtained as

$$u = \alpha_\zeta - v_{m,\zeta+1} + \hat{\rho}y_r^{(\zeta)} \tag{7.38}$$

$$\dot{\hat{\Theta}} = \Gamma\tau_\zeta - \sigma_2\Gamma\hat{\Theta} \tag{7.39}$$

where σ_2 is a positive design parameter and is chosen based on the σ-modification scheme proposed in [43]. A block diagram of the resulting closed-loop control systems is given in Figure 7.1.

7.3.3 Stability Analysis

We now analyze the designed controller and establish the stability of the closed-loop system and its tracking performance, as stated in the following theorem.

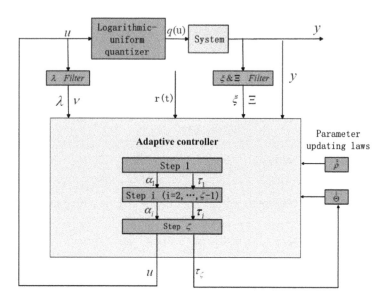

Figure 7.1 The closed-loop system

Theorem 7.1

Consider the closed-loop system consisting of the uncertain system (7.1) with the input signal u quantized by bounded quantizers, the filters (7.5)–(7.8), the control law (7.38) and parameter updating law (7.27), (7.39). All the signals of the closed-loop system are globally bounded and the tracking error $y - y_r$ will exponentially converge towards a set which is adjustable by choosing suitable design parameters.

Proof: We choose the Lyapunov function V_ζ as

$$V_\zeta = \sum_{i=1}^{\zeta} \frac{1}{2} z_i^2 + \frac{1}{2} \tilde{\Theta}^T \Gamma^{-1} \tilde{\Theta} + \frac{|b_m|}{2\gamma} \tilde{\rho}^2 \tag{7.40}$$

Then we get

$$\dot{V}_\zeta = \sum_{i=1}^{\zeta} z_i \dot{z}_i - \tilde{\Theta}^T \Gamma^{-1} \dot{\tilde{\Theta}} - \frac{b_m}{\gamma} \tilde{\rho} \dot{\hat{\rho}}$$

$$\leq -\sum_{i=1}^{\zeta} c_i z_i^2 + \frac{\sigma_1 |b_m|}{\gamma} \tilde{\rho} \hat{\rho} + \sigma_2 \tilde{\Theta}^T \hat{\Theta} + \sum_{i=1}^{\zeta} \frac{1}{4d_i} \epsilon^T \epsilon \tag{7.41}$$

Using Young's inequanlities $\tilde{\theta}^T \hat{\theta} \leq -\frac{1}{2}\tilde{\theta}^T\tilde{\theta} + \frac{1}{2}\theta^T\theta$ gives

$$\dot{V}_\zeta \leq -\sum_{i=1}^{\zeta} c_i z_i^2 - \frac{\sigma_1|b_m|}{2\gamma}\tilde{\rho}^2 + \frac{\sigma_1|b_m|}{2\gamma}\rho^2 - \frac{\sigma_2}{2}\tilde{\Theta}^T\tilde{\Theta}$$
$$+ \frac{\sigma_2}{2}\Theta^T\Theta + \sum_{i=1}^{\zeta} \frac{\alpha||P||\,||B||^2\lambda_{max}(P)\Delta_q^2}{4d_i\beta\lambda_{min}(P)} \tag{7.42}$$

Let $\mu = min\{2c_1,\ldots,2c_\rho,\sigma_1,\frac{\sigma_2}{\lambda_{max}(\Gamma^{-1})}\}$ where $\lambda_{max}(\Gamma^{-1})$ is the maximum eigenvalue of Γ^{-1}, and $\Delta = \frac{\sigma_1|b_m|}{2\gamma}\rho^2 + \frac{\sigma_2}{2}\Theta^T\Theta + \sum_{i=1}^{\zeta}\frac{\alpha||P||\,||B||^2\lambda_{max}(P)\Delta_q^2}{4d_i\beta\lambda_{min}(P)}$ which is bounded from Lemma 7.1. Then we can obtain

$$\dot{V}_\zeta \leq -\mu V_\zeta + \Delta \tag{7.43}$$

From (7.43), we have

$$\begin{aligned} V_\zeta(t) &\leq e^{-\mu t}V_\zeta(0) + (\Delta/\mu)(1 - e^{-\mu t}) \\ &\leq V_\zeta(0) + \Delta/\mu \end{aligned} \tag{7.44}$$

By (7.44), we get V_ζ is uniformly bounded. Thus z_i, $\hat{\Theta}$, and $\hat{\rho}$ are bounded. Since z_1 and y_r are bounded, y is also bounded. From (7.5) and (7.6), ξ and Ξ are bounded because A_0 is Hurwitz. Now we prove the boundedness of λ, then the boundedness of x follows from the boundedness of ϵ, ξ, Ξ, and λ. From (7.7), we obtain

$$\lambda_i = \frac{s^{i-1} + k_1 s^{i-2} + \cdots + k_{i-1}}{K(s)} u \tag{7.45}$$

where $i = 1,\ldots,n$, $K(s) = s^n + k_1 s^{n-1} + \cdots + k_n$. Meanwhile, the system model gives

$$\frac{d^n y}{dt^n} - \sum_{i=1}^{n} \frac{d^{n-i}}{dt^{n-i}}[\varphi_i(y) + \phi_i(y)^T\theta] = \sum_{i=0}^{m} b_i \frac{d^i}{dt^i}q(u) \tag{7.46}$$

Let $o = q(u) - u$, then o must be bounded. Noting that $\sum_{i=0}^{m} b_i \frac{d^i}{dt^i}q(u) = B(s)(u + o)$ and substituting (7.46) into (7.45), we get

$$\begin{aligned} \lambda_i &= \frac{s^{i-1} + k_1 s^{i-2} + \cdots + k_{i-1}}{B(s)K(s)} \\ &\quad \times \left(\frac{d^n y}{dt^n} - \sum_{i=1}^{n}\frac{d^{n-i}}{dt^{n-i}}[\varphi_i(y) + \phi_i(y)^T\theta] - o\right) \end{aligned} \tag{7.47}$$

Since $\varphi(y)$ and $\phi(y)$ are smooth, y and o are bounded, then from Assumption 7.2, $\lambda_1,\ldots,\lambda_{m+1}$ are bounded. By following a recursive analysis similar to [53], we can obtain that λ_i for all $i = m + 2,\ldots,n$ are bounded and therefore x is also bounded. Since the right hand side of (7.38) is a function of y, ξ, Ξ, and λ, then the control signal u is bounded. Therefore, the boundedness of all the closed-loop signals is guaranteed.

Moreover, from the definition of z_1, V_ζ, we obtain that the tracking error $e = z_1$ satisfies

$$\frac{1}{2}e^2 = \frac{1}{2}z_1^2 \le \frac{1}{2}\sum_{i=1}^{\zeta} z_i^2 \le V_\zeta(t)$$

$$\le e^{-\mu t}V_\zeta(0) + (\Delta/\mu)(1 - e^{-\mu t}) \tag{7.48}$$

So e^2 is bounded by a function that converges exponentially towards a compact set $\Omega = \{e|e^2 \le 2\Delta/\mu = 2 \times \frac{\frac{\sigma_1|b_m|}{2\gamma}\rho^2 + \frac{\sigma_2}{2}\Theta^T\Theta + \sum_{i=1}^{\zeta}\frac{\alpha||P|| \; ||B||^2\lambda_{max}(P)\Delta_q^2}{4d_i\beta\lambda_{min}(P)}}{\mu}\}$ at a rate of μ. As can be seen, the quantization error Δ_q and the vector Θ of all the unknown system parameters are both included in Ω, so they will affect the tracking error. But the size of Ω can be reduced by choosing suitable design parameters. Specifically, when the convergence speed μ, the parameters of the quantizer are set, the size of Ω can be reduced by decreasing $\lambda_{max}(\Gamma^{-1})$ and $\frac{\sigma_1}{\gamma}$ or increasing d_i. ■

7.3.4 Simulation Results

For simulation, the following system with a quantized input signal is considered:

$$\dot{x}_1 = x_2 + \theta y^3$$
$$\dot{x}_2 = b_0 q(u) \tag{7.49}$$

where $q(u)$ is the quantized input and θ, b_0 are the unknown parameters. Two cases are considered for the quantizer: hysteresis-uniform quantizer and logarithmic-uniform quantizer. The parameters of the hysteresis quantizer and the logarithmic quantizer are $d = 0.02$, $\delta = 0.2$, $u_i = 8.8$, respectively. Then the parameters $u_{th} = 11$ and $\varpi = 2.2$ are set for both cases. For the sake of simulation, we choose $\theta = 2$, $b_0 = 1$, $x_1(0) = 0$, $x_2(0) = 0.8$, and all the other initial conditions are set to be 0. We set the convergence rate $\mu = 0.5$, $c_1 = 100$, and $c_2 = 120$, so the size of the ultimate tracking error bound Λ is mainly determined by $\lambda_{max}(\Gamma^{-1})$, σ_1, σ_2, γ and d_1, d_2. For comparison, we choose the following two sets of parameters: (1). $\Gamma^{-1} = diag[10, 10]$, $\sigma_1 = 5$, $\sigma_2 = 5$, $\gamma = 0.1$, $d_1 = 2$, $d_2 = 1$; (2). $\Gamma^{-1} = diag[1, 1]$, $\sigma_1 = 0.5$, $\sigma_2 = 0.5$, $\gamma = 0.1$, $d_1 = 20$, $d_2 = 10$.

For the closed loop system with hysteresis-uniform quantizer, Figure 7.2 shows the state estimation performances, while Figure 7.3 shows the tracking performance of $y(t)$ and Figure 7.4 shows the input u and the quantized signal $q(u(t))$, respectively. On the other hand, for the system with logarithmic-uniform quantizer, Figures 7.5–7.7 illustrate the respective results.

By comparing the corresponding trajectories in the figures, it is observed that the ultimate tracking error bound can be reduced by decreasing $\lambda_{max}(\Gamma^{-1})$ and $\frac{\sigma_1}{\gamma}$ or increasing d_i, which is consistent with the established theoretical results.

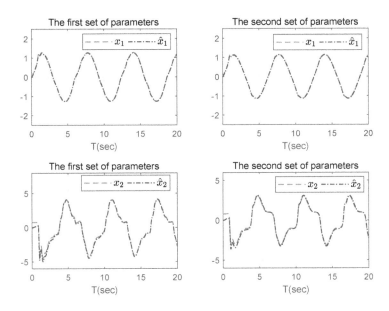

Figure 7.2 State estimation performance with Hysteresis-Uniform Quantizer

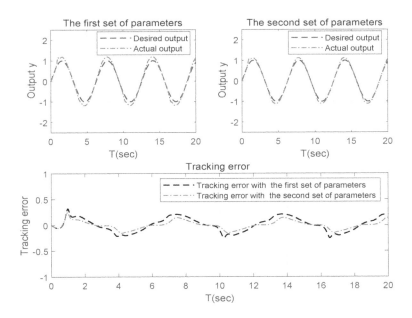

Figure 7.3 Tracking performance with Hysteresis-Uniform Quantizer

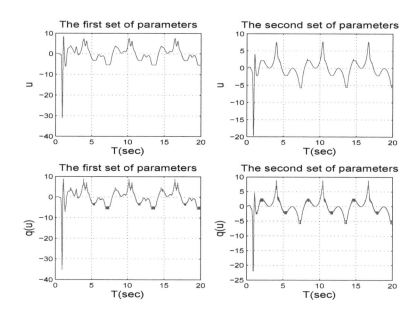

Figure 7.4 Control u and q(u) with Hysteresis-Uniform Quantizer

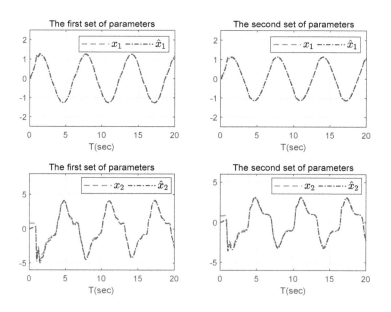

Figure 7.5 State estimation performance with Logarithmic-Uniform Quantizer

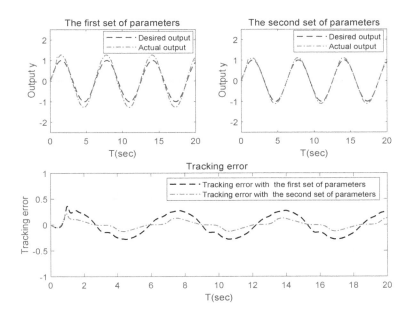

Figure 7.6 Tracking performance with Logarithmic-Uniform Quantizer

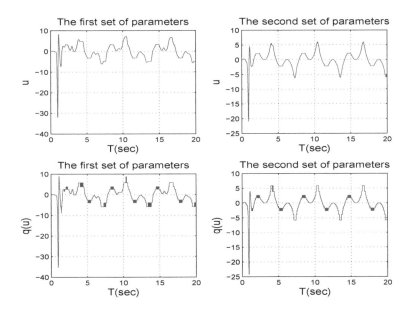

Figure 7.7 Control u and q(u) with Logarithmic-Uniform Quantizer

7.4 Controller Design with Sector-Bounded Quantizers

The above section presents an output-feedback control scheme for systems with bounded input quantization. However, the proposed scheme cannot be applied to sector-bounded quantizers whose quantization error cannot be assumed bounded in advance. The main reason lies in that, if the quantization errors are unbounded, the convergence property of the state estimation filters presented in Lemma 7.1 fails to hold. Therefore, new state estimation filter and then new controllers need to be designed for general sector-bounded quantizers. In the following, we will present a solution to this problem.

In this section, we present the controller design method for sector-bounded quantizers satisfying the property in (3.7). Typical examples include logarithmic quantizer in (3.2) and hysteresis-logarithmic quantizer in (3.4) in Chapter 3.

7.4.1 State Estimation Filters

Different form the observer designs in Section 7.3, the quantized signal $q(u)$, instead the input signal u, is utilized to design the state estimators. As a result, the filters are designed as follows:

$$\dot{\xi} = A_0\xi + ky + \Psi(y) \tag{7.50}$$

$$\dot{\Xi}^T = A_0\Xi^T + \Phi(y) \tag{7.51}$$

$$\dot{\lambda} = A_0\lambda + e_n q(u) \tag{7.52}$$

$$\nu_i = A_0^i\lambda, \quad i = 0, 1, \ldots, m, \tag{7.53}$$

where $e_i(i = 1, \ldots, n)$ is a row vector of n elements with the ith entry being 1 and the others being 0, $k = [k_1, \ldots, k_n]^T$ such that all the eigenvalues of $A_0 = A - ke_1^T$ are in the open left-half plane.

The state estimation filters are given as

$$\hat{x}(t) = \xi + \Xi^T\theta + \sum_{i=0}^{m} b_i\nu_i. \tag{7.54}$$

Then we get the derivative of $\hat{x}$ as

$$\dot{\hat{x}}(t) = A_0\hat{x} + ky + \Phi(y)\theta + \Psi(y) + \begin{pmatrix} 0 \\ b \end{pmatrix} q(u). \tag{7.55}$$

The state estimation error is defined as

$$\epsilon = x(t) - \hat{x}(t). \tag{7.56}$$

Lemma 7.2
For system (7.1), the proposed filters (7.50)–(7.52) guarantee that the state estimation errors converge exponentially to zero, regardless of the control signal.

Proof: From (7.1) and (7.54)–(7.56), the state estimation error satisfies

$$\dot{\epsilon} = A_0 \epsilon. \tag{7.57}$$

Since A_0 is Hurwitz, the state estimation error converge exponentially to zero, regardless of the control signal. ■

Remark 7.2 Note that (7.5)–(7.8) utilizes the input signal u to design the state estimator, which results in $\dot{\lambda} = A_0\lambda + e_n u$. Because $q(u) - u$ is always bounded for bounded quantizers regardless of the control signal u, the state estimation errors can be guaranteed bounded. However, this design method cannot be extended to the logarithmic (or hysteresis-logarithmic) quantizer because quantization errors cannot be assumed bounded. To solve this problem, we utilize the quantized input signal $q(u)$, which is also available, to design the state estimator, and the state estimation errors can be ensured to converge to zero.

7.4.2 Controller Design and Stability Analysis

With the new state estimation filters and following the control design procedures in Section 7.3, we can design the new controller in the following way:

$$u = -\frac{1}{\kappa}(\nu_{m,\zeta+1} - \hat{\rho}y_R^{(\zeta)} - \alpha_\zeta)\tanh(\frac{z_\zeta(\nu_{m,\zeta+1} - \hat{\rho}y_R^{(\zeta)} - \alpha_\zeta)}{\varepsilon})$$
$$- \frac{1}{\kappa}D\tanh(\frac{z_\zeta D}{\varepsilon}), \tag{7.58}$$

$$\dot{\hat{\Theta}} = \Gamma\tau_\zeta - \sigma_2\Gamma\hat{\Theta}, \tag{7.59}$$

where $0 < \kappa \leq 1 - \delta$, D is a constant satisfying $D \geq (1-\delta)d$, ε is a positive design constant, and σ_2 is a design parameter. The structure of the closed-loop systems obtained by the proposed control scheme is illustrated in Fig.7.8.

By analyzing the stability of the closed-loop system and the tracking performance, the following theorem is established.

Theorem 7.2
Consider the closed-loop system consisting of (7.1) where $q(u)$ is a sector bounded quantizer, the filters (7.50)–(7.52), the control law (7.58) and parameter updating law (7.27) and (7.59). All the closed-loop signals are globally bounded and the tracking error $y - y_r$ will exponentially converge towards a small region around zero which is adjustable by choosing suitable design parameters.

Proof: Firstly, the derivative of z_ζ is given as

$$\dot{z}_\zeta = q(u) + \nu_{m,\zeta+1} - k_\zeta\nu_{m,1} - \dot{\alpha}_{\zeta-1} - \hat{\rho}y_r^{(\zeta-1)} - \hat{\rho}y_R^{(\zeta)} \tag{7.60}$$

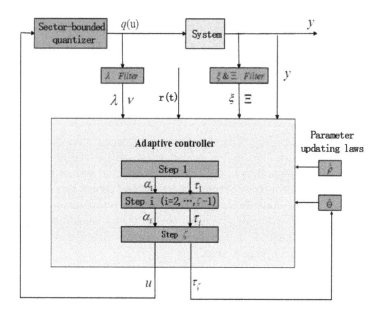

Figure 7.8 The closed-loop system

Now, we establish a property for $q(u)$. Note that it is sufficient to analyse the condition that $u > 0$, as the same analysis can be repeated for the case that $u \leq 0$. From (3.7), we know

$$u - \delta u - (1 - \delta)d \leq q(u) \leq u + \delta u + (1 - \delta)d. \tag{7.61}$$

Then for all $q(u)$, we have

$$u - \delta u - (1 - \delta)d \leq q(u) \leq u + \delta u + (1 - \delta)d. \tag{7.62}$$

This yields

$$q(u) = u + \eta \delta x + \eta_1 (1 - \delta)d = (1 + \eta \delta)u + \Delta, \tag{7.63}$$

where $\eta \in [-1, 1]$, $\eta_1 \in [-1, 1]$, and $\Delta = \eta_1(1 - \delta)d$. Substituting (7.63) into (7.60) gives

$$\dot{z}_\zeta = (1 + \eta \delta)u + \Delta + \nu_{m,\zeta+1} - k_\zeta \nu_{m,1} - \dot{\alpha}_{\zeta-1} - \hat{\rho} y_r^{(\zeta-1)} - \hat{\rho} y_R^{(\zeta)}. \tag{7.64}$$

We choose the Lyapunov function V_ζ as

$$V_\zeta = \sum_{i=1}^{\zeta} \frac{1}{2} z_i^2 + \frac{1}{2} \tilde{\Theta}^T \Gamma^{-1} \tilde{\Theta} + \frac{|b_m|}{2\gamma} \tilde{\rho}^2. \tag{7.65}$$

Then, we have

$$
\begin{aligned}
\dot{V}_\zeta =& \dot{V}_{\zeta-1} + z_\zeta((1+\eta\delta)u + \Delta + \nu_{m,\zeta+1} \\
& - k_\zeta\nu_{m,1} - \dot{\alpha}_{\zeta-1} - \hat{\rho}\dot{y}_r^{(\zeta-1)} - \hat{\rho}\dot{y}_R^{(\zeta)}) \\
\leq& -\sum_{i=1}^{\zeta} c_i z_i^2 + \frac{\sigma_1|b_m|}{\gamma}\tilde{\rho}\hat{\rho} + \sigma_2\tilde{\Theta}^T\hat{\Theta} + \sum_{i=1}^{\zeta}\frac{1}{4d_i}\epsilon^T\epsilon \\
& + z_\zeta((1+\eta\delta)u + \Delta + \nu_{m,\zeta+1} - \hat{\rho}\dot{y}_R^{(\zeta)} - \alpha_\zeta).
\end{aligned}
\tag{7.66}
$$

From (7.58), we have $z_\zeta u \leq 0$. Since $\eta \in [-1,1]$, $(1+\eta\delta) \geq \kappa = 1 - \delta > 0$. As a result, $z_\zeta(1+\eta\delta)u \leq \kappa z_\zeta u$. Note that the hyperbolic tangent function $\tanh(\cdot)$ has the following property

$$
0 \leq |\varrho| - \varrho\tanh(\frac{\varrho}{\varepsilon}) \leq 0.2785\varepsilon,
\tag{7.67}
$$

where $\varepsilon > 0$ and $\varrho \in \mathbb{R}$. Therefore, substituting (7.58) into (7.66) gives

$$
\dot{V}_\zeta \leq -\sum_{i=1}^{\zeta} c_i z_i^2 + \frac{\sigma_1|b_m|}{\gamma}\tilde{\rho}\hat{\rho} + \sigma_2\tilde{\Theta}^T\hat{\Theta} + \sum_{i=1}^{\zeta}\frac{1}{4d_i}\epsilon^T\epsilon + 0.557\varepsilon.
\tag{7.68}
$$

Using Young's inequalities $\tilde{\theta}^T\hat{\theta} \leq -\frac{1}{2}\tilde{\theta}^T\tilde{\theta} + \frac{1}{2}\theta^T\theta$ gives

$$
\begin{aligned}
\dot{V}_\zeta \leq& -\sum_{i=1}^{\zeta} c_i z_i^2 - \frac{\sigma_1|b_m|}{2\gamma}\tilde{\rho}^2 + \frac{\sigma_1|b_m|}{2\gamma}\rho^2 \\
& - \frac{\sigma_2}{2}\tilde{\Theta}^T\tilde{\Theta} + \frac{\sigma_2}{2}\Theta^T\Theta + \sum_{i=1}^{\zeta}\frac{1}{4d_i}\epsilon^T\epsilon + 0.557\varepsilon.
\end{aligned}
\tag{7.69}
$$

Let

$$
\mu = min\{2c_1,\ldots,2c_\zeta,\sigma_1,\frac{\sigma_2}{\lambda_{max}(\Gamma^{-1})}\}
\tag{7.70}
$$

where $\lambda_{max}(\Gamma^{-1})$ is the maximum eigenvalue of Γ^{-1} and

$$
\Delta = \frac{\sigma_1|b_m|}{2\gamma}\rho^2 + \frac{\sigma_2}{2}\Theta^T\Theta + \sum_{i=1}^{\zeta}\frac{1}{4d_i}\epsilon^T\epsilon + 0.557\varepsilon
\tag{7.71}
$$

which is bounded. Then we can obtain

$$
\dot{V}_\zeta \leq -\mu V_\zeta + \Delta.
\tag{7.72}
$$

Following the same analyses as in Theorem 7.1, all the closed-loop signals are globally bounded. Moreover, from the definition of z_1, V_ζ, we obtain that the tracking error $e = z_1$ satisfies

$$
\frac{1}{2}e^2 = \frac{1}{2}z_1^2 \leq \frac{1}{2}\sum_{i=1}^{\zeta} z_i^2 \leq V_\zeta(t) \leq e^{-\mu t}V_\zeta(0) + (\Delta/\mu)(1 - e^{-\mu t}).
\tag{7.73}
$$

So e^2 is bounded by a function that converges exponentially towards a compact set $\Omega = \{e|e^2 \leq 2\Delta/\mu = 2 \times \frac{\frac{\sigma_1|b_m|}{2\gamma}\rho^2 + \frac{\sigma_2}{2}\Theta^T\Theta + \sum_{i=1}^{\varsigma}\frac{1}{4d_i}\epsilon^T\epsilon + 0.557\varepsilon}{\mu}\}$ at a rate of μ. As can be seen, the size of Ω can be reduced by choosing suitable design parameters. Specifically, when the convergence speed μ and the parameters of the utilized quantizer are set, the size of Ω can be reduced by decreasing $\lambda_{max}(\Gamma^{-1})$, $\frac{\sigma_1}{\gamma}$ and ε or increasing d_i. ∎

Remark 7.3   One key technique to handle the quantization error is to find a way so that $z_\varsigma u < 0$ is guaranteed, as this together with the sector bound property of the quantizers gives the crucial property $z_\varsigma q(u) = z_\varsigma(1+\eta\delta)u \leq \kappa z_\varsigma u$. This is achieved by employing the hyperbolic tangent function $\tanh(\cdot)$ in the controller design. Besides, the designed controller depends on the quantizer parameter δ and d (see the definition of κ and D below (7.59)). In summary, the effects of the quantization error on the closed-loop system are effectively handled in a robust way.

7.4.3   Simulation

Now, we illustrate an application of our proposed scheme in controlling a ship. Following [103], the dynamics of a ship can be modeled as

$$\ddot{y} + \Phi\dot{y} + b_0(My^3 + Ly) = b_0 q(u) \tag{7.74}$$

where y is the course angular velocity, $q(u)$ is the quantized input, $b_0 \neq 0$, M and L are unknown constants related to the hydrodynamic coefficients and the mass of the ship. Let $x_1 = y$, $x_2 = \dot{y} + \Phi y$, $\theta = [\Phi, b_0 M, b_0 L]^T$, $b = [0, b_0]^T$, and $\Psi(y) = \begin{pmatrix} -y & 0 & 0 \\ 0 & -y^3 & -y \end{pmatrix}$, then system (7.74) can be transformed into

$$\begin{pmatrix} \dot{x}_1 \\ \dot{x}_2 \end{pmatrix} = \begin{pmatrix} 0 & 1 \\ 0 & 0 \end{pmatrix} \begin{pmatrix} x_1 \\ x_2 \end{pmatrix} + \Psi(y)\theta + bq(u) \tag{7.75}$$

The control objective is to make the output $y = x_1(t)$ track the reference signal which is the output of a reference system $\ddot{y} + 2\dot{y} + y = 1.5\sin(0.6t)$. In this case, we test both the hysteresis-logarithmic and logarithmic quantizers, and their parameters are set as $d = 0.02$, $\delta = 0.2$. Similar to [103], for simulation studies, the system parameters are set as $\Phi = 0.2$, $b_0 = 1.85$, $M = 0.12$, $L = 0.28$, but they are totally unknown in controller design. In the simulation, we choose $k_1 = 3$, $k_2 = 2$, $y(0) = 0.7$rad/s, $\dot{y}(0) = 0$rad/s^2, and all the other initial conditions are 0. Besides, we set the convergence rate $\mu = 0.1$, $c_1 = c_2 = 2$, $\kappa = 0.8$ and $D = 0.2$. For comparisons, two sets of parameters are considered: 1). $\Gamma^{-1} = diag[5, 5]$, $\sigma_1 = \sigma_2 = 0.5$, $\gamma = 2$, $d_1 = 2$, $d_2 = 1$, $\varepsilon = 20$; 2). $\Gamma^{-1} = diag[1, 1]$, $\sigma_1 = \sigma_2 = 0.1$, $\gamma = 2$, $d_1 = 4$, $d_2 = 2$, $\varepsilon = 10$.

   For the closed-loop system with the hysteresis-logarithmic quantizer, Figure 7.9 shows the tracking performance of $y(t)$ and Figure 7.10 shows the input u and the quantized signal $q(u(t))$. On the other hand, Figures 7.11–7.12 illustrate the respective results for the system with the logarithmic qunatizer. It is observed that the ultimate tracking error bound can be reduced by decreasing $\lambda_{max}(\Gamma^{-1})$ and $\frac{\sigma_1}{\gamma}$ or increasing d_i, which is consistent with the established theoretical results.

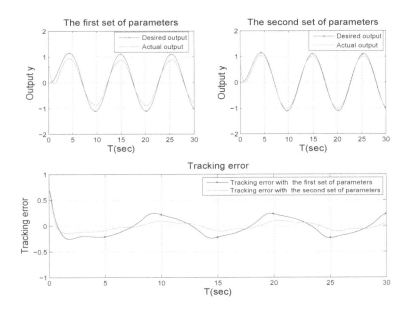

Figure 7.9 Tracking performance with hysteresis-logarithmic quantizer

7.5 Notes

In this chapter, adaptive backstepping output-feedback control schemes are designed for a class of uncertain nonlinear systems with input quantization. Two types of quantizers are considered here, i.e. bounded-quantizers and sector-bounded quantizers. It is shown that the designed adaptive controllers ensure global boundedness of all the signals in the closed-loop system and enables the tracking error to exponentially converge towards a compact set which is adjustable.

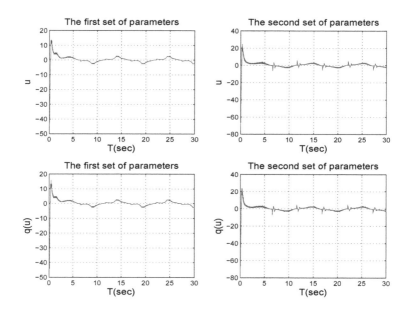

Figure 7.10 Control u and q(u) with hysteresis-logarithmic quantizer

Acknowledgment

Reprinted from Copyright (2021), with permission from Elsevier. Lantao Xing, Changyun Wen, Yang Zhu, Hongye Su, Zhitao Liu, "Output Feedback Control for Uncertain Nonlinear Systems with Input Quantization", *Automatica*, vol. 65, pp. 191–202, 2016.

Reprinted from Copyright (2021), with permission from John Wiley and Sons. Lantao Xing, Changyun Wen, Hongye Su, Guanyu Lai, Zhitao Liu, "Robust Adaptive Output Feedback Control for Uncertain Nonlinear Systems with Quantized Input", *International Journal of Robust and Nonlinear Control*, vol. 27, pp.1999–2016, 2017.

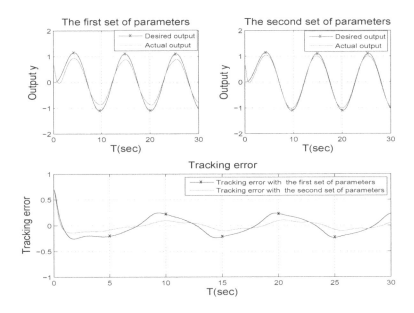

Figure 7.11 Tracking performance with logarithmic quantizer

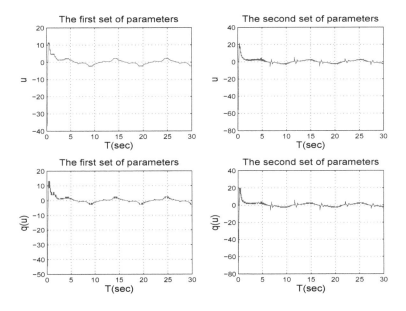

Figure 7.12 Control u and q(u) with logarithmic quantizer

STATE
QUANTIZATION
COMPENSATION

Chapter 8

Adaptive Control of Systems with Bounded State Quantization

This chapter investigates the stabilization problem for uncertain nonlinear systems with quantized states. All states in the system are quantized by bounded quantizers, such as the uniform quantizer, the hysteresis-uniform quantizer and the logarithmic-uniform quantizer in Chapter 3. Adaptive backstepping based control algorithms and a new approach to stability analysis are developed by constructing a new compensation scheme for effects of the state quantization and to handle discontinuity resulted from state quantization. Besides showing the ultimate boundedness of the system, the stabilization error performances are also established and can be improved by appropriately adjusting design parameters. Simulation results illustrate the effectiveness of the proposed scheme.

8.1 Introduction

Feedback control of systems with state quantization has attracted growing interest lately in [60, 63, 67]. For a control system with state quantization, the state measurements are processed by quantizers, which are discontinuous maps from continuous spaces to finite sets. Such discontinuous property may make the control design and stability analysis difficult. Note that in [60, 63, 67], the systems considered are completely known. Uncertainties and nonlinearities always exist in many practical systems. Thus, it is more reasonable to consider controller design for uncertain nonlinear systems. Although adaptive control of uncertain systems has received considerable interest and has been widely investigated, there are still limited works devoted to adaptive control with state quantization. It is noted that adaptive control schemes for linear systems with state quantization have been reported only in [3, 99]. In [99], a supervisory control scheme for uncertain linear systems with quantized measurements has been proposed. While in [3], the adaptive control of linear systems with quantized measurements and bounded disturbances has been addressed.

Research on adaptive control of input quantization using the backstepping technique has also received great attention, see for examples, [127, 142, 147]. The major difficulty to deal with the state quantization using the backstepping technique is that the backstepping technique requires differentiating virtual controls and in turn the states by applying the chain rule. If the states are quantized, they become discontinuous, and therefore it is difficult to analyze the resulting control system with the current backstepping-based approaches. In this chapter, we provide a solution to this problem by developing a new adaptive controller and a new approach to stability analysis. The quantizers considered in this chapter are static and satisfy a bounded condition. Typical examples include a uniform quantizer, a hysteresis uniform quantizer, and a combination of logarithmic and uniform quantizer.

The main contributions are summarized as follows.

■ Note that the existing backstepping design procedure requires recursive differentiation of the virtual controls and in turn the states. Thus these variables should be sufficiently smooth. In this chapter, we make a significant modification of standard adaptive backstepping in the way that the virtual control laws only use the partial derivatives which are constants and depend on the control design parameters. It is this new technique that enables us to successfully overcome the difficulties caused by the discontinuity and the bounded uncertainties resulted from state quantization.

■ A new method is proposed to compensate for the effects of the state quantization. To handle fully unknown parameters, new parameter updating laws, which do not require any information on such unknown parameters including the knowledge on their bounds, are developed. Thus a new adaptive control scheme is developed to achieve desired stability and performance for a class of nonlinear systems. More specifically, 1) the stability in the sense of ultimate boundedness is achieved by choosing suitable design parameters; 2) The upper bound of ultimate stabilization error can be decreased by tuning some design parameters.

8.2 Problem Statement

8.2.1 System Model

We consider a class of nonlinear uncertain systems described as follows

$$x^{(n)}(t) \;=\; u(t) + \psi\left(x, \dot{x}, \ldots, x^{(n-1)}\right) + \phi^T\left(x, \dot{x}, \ldots, x^{(n-1)}\right)\theta \quad (8.1)$$

where $(x(t), \dot{x}(t), \ldots, x^{(n-1)}(t))^T \in \Re^n$ and $u(t) \in \Re^1$ are the states and input of the system respectively, $\psi \in \Re^1$ and $\phi \in \Re^r$ are known nonlinear functions, $\theta \in \Re^r$ is a vector of unknown constant parameters, $q(\cdot)$ has an infinite level. Such a class of nonlinear systems has been addressed in many references, such as [51, 71, 148, 152]. It was noted in [46, 51, 71] that various practically important systems can be transformed to this structure.

In this chapter, only quantized states $(q(x), q(\dot{x}), \ldots, q(x^{(n-1)}))$ are measured. The feedback controller $u(t)$ in (8.1) only uses the quantized states, which is given by

$$u \;=\; u(q(x), q(\dot{x}), \ldots, q(x^{(n-1)})). \quad (8.2)$$

For the development of control laws, the following assumption is made.

Assumption 8.1 *The functions ψ and ϕ satisfy the global Lipschitz continuity condition such that*

$$|\psi(y_1) - \psi(y_2)| \;\leq\; L_\psi \| y_1 - y_2 \| \quad (8.3)$$
$$\| \phi(y_1) - \phi(y_2) \| \;\leq\; L_\phi \| y_1 - y_2 \| \quad (8.4)$$

where L_ψ and L_ϕ are constants, $y_1, y_2 \in \Re^n$ are real vectors. The norm $\| \cdot \|$ is defined as $\| y \| = (\sum_{j=1}^m y_j^2)^{1/2}$ for a vector $y = [y_1, \ldots, y_m]^T$. $|\cdot|$ denotes the absolute value of a scalar.

Assumption 8.2 *For the closed-loop nonlinear uncertain system (8.1)–(8.2), it is assumed that its solution exists and is unique.*

Note that only quantized states $\left(q(x), q(\dot{x}), \ldots, q(x^{(n-1)})\right)$ are measurable and available for control design, instead of the states $(x, \dot{x}, \ldots, x^{(n-1)})$. Similar assumptions are also made in the area, for instances [34, 60, 145–147]. As illustrated in the simulation studies, the designed adaptive controller with parameter estimator guarantees the existence and uniqueness of the solution.

The control objective is to design an adaptive controller u (8.2) for system (8.1) by utilizing only quantized states $(q(x), q(\dot{x}), \ldots, q(x^{(n-1)}))$ such that all the signals in the closed-loop system are globally uniformly bounded.

8.2.2 Bounded Quantizer

The quantizer $q(x)$ considered in this chapter has the following property similar to (3.6).

$$|q(x) - x| \leq \delta \qquad (8.5)$$

where $\delta > 0$ is the quantization bound. It is shown in Chapter 3 that the uniform quantizer in (3.1), the hysteresis-uniform quantizer in (3.3), and the logarithmic-uniform quantizer in (3.5) satisfy the property (8.5).

Remark 8.1 Note that the quantization parameters are not required to be known for the control design, such as l for uniform quantizer in (3.1), l and h for hysteresis-uniform quantizer in (3.3), l, ρ and x_{th} for logarithmic-uniform quantizer in (3.5). The uniform quantizer and the hysteresis-uniform have uniformly spaced quantization levels which is optimal for uniformly distributed signals. Compared with the uniform quantizer, the hysteresis-uniform quantizer has additional quantization levels, which are used to avoid chattering. A logarithmic-uniform quantizer can minimize the average rate of communication instances. Detailed discussions can be found in Chapter 3.

8.3 Adaptive Backstepping Control

In order to design the controller using backstepping technique, system (8.1) is rewritten in the following form

$$
\begin{aligned}
\dot{x}_1 &= x_2 \\
\dot{x}_i &= x_{i+1}, \quad i = 1, \dots, n-1 \\
\dot{x}_n &= u(t) + \psi(x_1, \dots, x_n) + \theta^T \phi(x_1, \dots, x_n)
\end{aligned} \qquad (8.6)
$$

where $x_1 = x$, $x_i = x^{(i-1)}$, $i = 2, 3, \dots, n$. The system states $(x_1, x_2 \dots, x_n) \in \Re$. are quantized by a quantizer satisfying the property in (8.5). Only the measured quantized states $q(x_i)$, $i = 1, \dots, n$ are available.

8.3.1 States are Not Quantized

If states x_i, $i = 1, 2, \dots, n$ are not quantized, we begin by introducing the change of coordinates

$$
\begin{aligned}
z_1(x_1) &= x_1 & (8.7) \\
z_i(x_1, \dots, x_i) &= x_i - \alpha_{i-1}, \ i = 2, 3, \dots, n & (8.8)
\end{aligned}
$$

where α_{i-1} is the virtual control function of $(x_1, \dots, x_{i-1})$ and will be determined at the ith step.

• *Step i ($i = 1, \ldots, n - 1$):* Following the standard backstepping design technique in [53], we choose

$$\alpha_1(x_1) = -c_1 z_1(x_1) \tag{8.9}$$

$$\alpha_i(x_1, \ldots, x_i) = -c_i z_i - z_{i-1} - \dot{\alpha}_{i-1}$$

$$= -c_i z_i - z_{i-1} + \sum_{k=1}^{i-1} \frac{\partial \alpha_{i-1}}{\partial x_k} x_{k+1}, \quad i = 2, \ldots, n - 1 \tag{8.10}$$

where $c_i, i = 2, \ldots, n - 1$ are positive design parameters and $\frac{\partial \alpha_{i-1}}{\partial x_k}$ are constants which depend on $c_1, \ldots, c_{i-1}$. For examples,

$$\frac{\partial \alpha_1}{\partial x_1} = \frac{\partial \alpha_1}{\partial z_1} \frac{\partial z_1}{\partial x_1} = -c_1 \tag{8.11}$$

$$\frac{\partial \alpha_2}{\partial x_1} = -c_2 \frac{\partial z_2}{\partial x_1} - \frac{\partial z_1}{\partial x_1} = -c_2 c_1 - 1 \tag{8.12}$$

$$\frac{\partial \alpha_2}{\partial x_2} = -c_2 \frac{\partial z_2}{\partial x_2} + \frac{\partial \alpha_1}{\partial x_1} = -c_2 - c_1. \tag{8.13}$$

Considering the Lyapunov function

$$V_{n-1} = \sum_{j=1}^{n-1} \frac{1}{2} z_j^2 \tag{8.14}$$

then the derivative is given as

$$\dot{V}_{n-1} = -\sum_{j=1}^{n-1} c_j z_j^2 + z_{n-1} z_n \tag{8.15}$$

• *Step n:* If states x_i, $i = 1, 2, \ldots, n$ are not quantized, the virtual control $\alpha_n(x_1, \ldots, x_n)$ is designed as

$$\alpha_n(x_1, \ldots, x_n) = -c_n z_n - z_{n-1} - \dot{\alpha}_n$$

$$= -c_n z_n - z_{n-1} + \sum_{k=1}^{n-1} \frac{\partial \alpha_{n-1}}{\partial x_k} x_{k+1} \tag{8.16}$$

where c_n is a positive design parameter and $\hat{\theta}$ is the estimate of θ. The final control $u(t)$ is chosen as

$$u(t) = \alpha_n - \psi(x_1, \ldots, x_n) - \hat{\theta}^T \phi(x_1, \ldots, x_n) \tag{8.17}$$

$$\dot{\hat{\theta}} = \Gamma \phi(x_1, \ldots, x_n) z_n, \tag{8.18}$$

where Γ is a positive definite matrix and $\hat{\theta}$ is the estimate of θ.

Considering the Lyapunov function

$$V = \sum_{j=1}^{n} \frac{1}{2} z_j^2 + \frac{1}{2} \tilde{\theta}^T \Gamma^{-1} \tilde{\theta} \tag{8.19}$$

where $\tilde{\theta} = \theta - \hat{\theta}$, then the derivative is given as

$$
\begin{aligned}
\dot{V} &= -\sum_{j=1}^{n-1} c_j z_j^2 + z_n \left(\alpha_n - \dot{\alpha}_{n-1} + z_{n-1} \right) - \tilde{\theta}^T \Gamma^{-1} \dot{\hat{\theta}} \\
&= -\sum_{j=1}^{n} c_j z_j^2 + \tilde{\theta}^T \Gamma^{-1} \left(\Gamma \phi z_n - \dot{\hat{\theta}} \right) \\
&= -\sum_{j=1}^{n} c_j z_j^2
\end{aligned}
\tag{8.20}
$$

Thus, we conclude that the closed-loop system without state quantization is globally asymptotically stable and the desired convergence property $\lim_{t\to\infty} z_i(t) = 0$ follows from LaSalle-Yoshizawa theorem in [53].

8.3.2 States are Quantized

When states x_i, $i = 1, 2, \ldots, n$ are quantized with quantizers $q(x_i)$, the adaptive controllers and the parameter updating law are chosen as

$$
\begin{aligned}
u(t) &= \bar{\alpha}_n - \bar{\psi} - \hat{\theta}^T \bar{\phi} \\
&= -c_n \bar{z}_n - \bar{z}_{n-1} - \psi \left(q(x_1), \ldots, q(x_n) \right) \\
&\quad -\hat{\theta}^T \phi \left(q(x_1), \ldots, q(x_n) \right) + \sum_{k=1}^{n-1} \frac{\partial \alpha_{n-1}}{\partial x_k} q(x_{k+1})
\end{aligned}
\tag{8.21}
$$

$$
\begin{aligned}
\dot{\hat{\theta}} &= \Gamma \bar{\phi} \bar{z}_n - \Gamma k_\theta (\hat{\theta} - \theta_0) \\
&= \Gamma \phi (q(x_1), \ldots, q(x_n)) \bar{z}_n - \Gamma k_\theta (\hat{\theta} - \theta_0)
\end{aligned}
\tag{8.22}
$$

$$
\bar{z}_1 = q(x_1)
\tag{8.23}
$$

$$
\bar{z}_i = q(x_i) - \bar{\alpha}_{i-1}
\tag{8.24}
$$

$$
\bar{\alpha}_1 = -c_1 \bar{z}_i
\tag{8.25}
$$

$$
\bar{\alpha}_i = -c_i \bar{z}_i - \bar{z}_{i-1} + \sum_{k=1}^{i-1} \frac{\partial \alpha_{i-1}}{\partial x_k} q(x_{k+1}), \ i = 2, \ldots, n
\tag{8.26}
$$

where c_i, k_θ, and θ_0 are positive parameters and Γ is a positive definite matrix.

Remark 8.2 Note that an additional term in the form of $-\Gamma k_\theta (\hat{\theta} - \theta_0)$ is introduced in the parameter estimator (8.22). It will be observed from subsequent stability analysis that by adopting such modification, the following property

$$
k_\theta \tilde{\theta} (\hat{\theta} - \theta_0) \leq -\frac{1}{2} k_\theta \| \tilde{\theta} \|^2 + \frac{1}{2} k_\theta \| (\theta - \theta_0) \|^2
\tag{8.27}
$$

can be obtained which is helpful to guarantee the closed-loop system stability. Unlike [147], no prior information about the bound of unknown parameter vector is required.

Remark 8.3   For the system with quantized states, the state x_i is not available and only the quantized state $q(x_i)$ can be used in the designed controller. If we follow the standard backstepping controller in Section 8.3.1, the virtual control $\bar{\alpha}_i$ should be like $-c_i \bar{z}_i - \bar{z}_{i-1} + \dot{\bar{\alpha}}_{i-1}$. Note that the quantized state $(q(x_1), q(x_2) \ldots q(x_{i-1}))$ is used in the virtual control $\bar{\alpha}_{i-1}$ which results in that the derivative of $\bar{\alpha}_{i-1}$ does not exist and thus is unable to be used in the backstepping virtual control.

Remark 8.4   Note that the final control u in (8.21) and the parameter updating law in (8.22) utilize only the measured quantized states $q(x_i)$, $i = 1, \ldots, n$. One vitally important technique adopted in this chapter is to use the partial derivatives $\frac{\partial \alpha_{i-1}}{\partial x_k} (i = 2, 3, \ldots, n, \ k = 1, \ldots, i-1)$ in the final control u in (8.21) and the function $\bar{\alpha}_i$ in (8.26), which cancels the effects caused by the previous virtual control $\bar{\alpha}_{i-1}$ in the stability analysis. Note that, as illustrated in the calculations (8.11)– (8.13), the partial derivatives $\frac{\partial \alpha_{i-1}}{\partial x_k} (i = 2, 3, \ldots, \ n, k = 1, \ldots, i-1)$ are constants and depend on the control gains $(c_1, \ldots, c_{i-1})$ chosen in each recursive step.

Remark 8.5   The change of coordinates z_i in (8.8) and the virtual control functions α_i in (8.10) are only used in the Lyapunov stability analysis, since the state x_i for $i = 1, \ldots, n$ is not used in the final controller and parameter estimator designed.

In order to ensure the boundedness of all signals, we first establish some preliminary results as stated in the following lemmas.

Lemma 8.1
The effects of state quantization are bounded as follows:

$$|\psi(q(x_1), \ldots, q(x_n)) - \psi(x_1, \ldots, x_n)| \ \leq \ \Delta_\psi \qquad (8.28)$$
$$\| \phi(q(x_1), \ldots, q(x_n)) - \phi(x_1, \ldots, x_n) \| \ \leq \ \Delta_\phi \qquad (8.29)$$
$$|z_i(q(x_1), \ldots, q(x_i)) - z_i(x_1, \ldots, x_i)| \ \leq \ \Delta_{z_i} \qquad (8.30)$$
$$|\alpha_i(q(x_1), \ldots, q(x_i)) - \alpha_i(x_1, \ldots, x_i)| \ \leq \ \Delta_{\alpha_i} \qquad (8.31)$$

where $i = 1, \ldots, n$, Δ_ψ and Δ_ϕ are positive constants which depend on the quantization bound δ and Lipschitz constants L_ψ and L_ϕ respectively. Δ_{z_i} is positive which depends on the quantization bound δ and control design parameters $(c_1, \ldots, c_{i-1})$, Δ_{α_i} is a positive constant which depends on the quantization bound δ and control design parameters $(c_1, \ldots, c_i)$.

Proof 8.1   Using the property of quantizer in (8.5), we have

$$|q(x_i) - x_i| \ \leq \ \delta. \qquad (8.32)$$

With the Lipschitz continuous conditions for ψ and ϕ in (8.3) and (8.4) in

Assumption 8.2, the following bounded conditions are obtained.

$$
\begin{aligned}
&|\psi\left(q(x_1),\ldots,q(x_n)\right) - \psi(x_1,\ldots,x_n)| \\
\leq\ & L_\psi \left\| \left(q(x_1),\ldots,q(x_n)\right) - (x_1,\ldots,x_n) \right\| \\
\leq\ & L_\psi \left\| (\delta,\ldots,\delta) \right\| = L_\psi \sqrt{n}\delta = \Delta_\psi
\end{aligned}
\tag{8.33}
$$

$$
\begin{aligned}
&\left\| \phi\left(q(x_1),\ldots,q(x_n)\right) - \phi(x_1,\ldots,x_n) \right\| \\
\leq\ & L_\phi \left\| \left(q(x_1),\ldots,q(x_n)\right) - (x_1,\ldots,x_n) \right\| \\
\leq\ & L_\phi \left\| (\delta,\ldots,\delta) \right\| = L_\phi \sqrt{n}\delta = \Delta_\phi
\end{aligned}
\tag{8.34}
$$

From (8.9)-(8.17), and (8.21)-(8.26), it is shown that

$$
\begin{aligned}
|\bar{z}_1 - z_1(x_1)| &= |z_1\left(q(x_1)\right) - z_1(x_1)| \\
&= |q(x_1) - x_1| \leq \delta \triangleq \Delta_{z_1}
\end{aligned}
\tag{8.35}
$$

$$
\begin{aligned}
|\bar{\alpha}_1 - \alpha_1(x_1)| &= |\alpha_1\left(q(x_1)\right) - \alpha_1(x_1)| \\
&= |-c_1(\bar{z}_1 - z_1)| \leq c_1\delta \triangleq \Delta_{\alpha_1}
\end{aligned}
\tag{8.36}
$$

$$
\begin{aligned}
|\bar{z}_2 - z_2(x_1,x_2)| &= |z_2\left(q(x_1),q(x_2)\right) - z_2(x_1,x_2)| \\
&= |q(x_2) - \bar{\alpha}_1 - (x_2 - \alpha_1)| \leq \delta + \Delta_{\alpha_1} \triangleq \Delta_{z_2}
\end{aligned}
\tag{8.37}
$$

$$
\begin{aligned}
|\bar{\alpha}_2 - \alpha_2(x_1,x_2)| &= |\alpha_2\left(q(x_1),q(x_2)\right) - \alpha_2(x_1,x_2)| \\
&= \left| -c_2(\bar{z}_2 - z_2) - (\bar{z}_1 - z_1) + \frac{\partial\alpha_1}{\partial x_1}(q(x_1) - x_1) \right| \\
&\leq c_2\Delta_{z_2} + \Delta_{z_1} + \left|\frac{\partial\alpha_1}{\partial x_1}\right|\delta \triangleq \Delta_{\alpha_2}
\end{aligned}
\tag{8.38}
$$

Following the same procedure based on z_i in (8.8), α_i in (8.10), $\bar{z}_i$ in (8.24), $\bar{\alpha}_i$ in (8.26), we have

$$
\begin{aligned}
|\bar{z}_i - z_i(x_1,\ldots,x_i)| &= |z_i\left((q(x_1),\ldots,q(x_i))\right) - z_i(x_1,\ldots,x_i)| \\
&\leq |(q(x_i) - x_i) - (\bar{\alpha}_{i-1} - \alpha_{i-1})| \\
&\leq \delta + \Delta_{\alpha_{i-1}} \triangleq \Delta_{z_i}
\end{aligned}
\tag{8.39}
$$

$$
\begin{aligned}
|\bar{\alpha}_i - \alpha_i(x_1,\ldots,x_i)| &= |\alpha_i\left((q(x_1),\ldots,q(x_i))\right) - \alpha_i(x_1,\ldots,x_i)| \\
&\leq |-c_i(z_i - \bar{z}_i) - (\bar{z}_{i-1} - z_{i-1})| \\
&\quad + \left|\sum_{k=1}^{i-1} \frac{\partial\alpha_{i-1}}{\partial x_k}\left(q_{k+1}(x_{k+1}) - x_{k+1}\right)\right| \\
&\leq c_i\Delta_{z_i} + \Delta_{z_{i-1}} + \sum_{k=1}^{i-1}\left|\frac{\partial\alpha_{i-1}}{\partial x_k}\right|\delta \triangleq \Delta_{\alpha_i}
\end{aligned}
\tag{8.40}
$$

Lemma 8.2

The states $(x_1,x_2,\ldots,x_n)$ satisfy the following inequality,

$$
\left\| (x_1,\ldots,x_n) \right\| \leq L_x \left\| (z_1,\ldots,z_n) \right\|
\tag{8.41}
$$

where L_x is a positive constant which depends on the control design parameters $(c_1, \ldots, c_{n-1})$.

Proof 8.2 From the definitions z_i in (8.7)–(8.8) and the virtual control designs α_i in (8.9)–(8.16), it is shown that

$$|x_1| = |z_1| \tag{8.42}$$

$$|\alpha_1| \leq c_1|z_1| \triangleq L_{\alpha_1}|z_1| \tag{8.43}$$

$$|x_2| \leq |z_2 + \alpha_1| \leq |z_2| + L_{\alpha_1}|z_1|$$
$$\leq \sqrt{2}\max\{1, L_{\alpha_1}\} \, \| (z_1, z_2) \| \triangleq L_{x2} \, \| (z_1, z_2) \| \tag{8.44}$$

$$|\alpha_2| \leq c_2|z_2| + \left| \frac{\partial \alpha_1}{\partial x_1} x_2 \right|$$
$$\leq (c_2 + c_1 L_{x2}) \, \| (z_1, z_2) \| \triangleq L_{\alpha_2} \, \| (z_1, z_2) \| \tag{8.45}$$

where L_{α_1} depends on c_1, L_{α_2} depends on c_1, c_2, and L_{x_2} depends on c_1. Following the similar procedure, we have

$$|x_i| \leq |z_i + \alpha_{i-1}|$$
$$\leq |z_i| + L_{\alpha_{i-1}} \, \| (z_1, \ldots, z_{i-1}) \|$$
$$\leq (1 + L_{\alpha_{i-1}}) \, \| (z_1, \ldots, z_{i-1}) \|$$
$$\triangleq L_{x_i} \, \| (z_1, z_2, \ldots, z_i) \| \tag{8.46}$$

$$|\alpha_i| \leq c_i|z_i| + \left| \sum_{j=1}^{i-1} \frac{\partial \alpha_{i-1}}{\partial x_j} x_{j+1} \right|$$
$$\leq \left(c_i + \left| \sum_{j=1}^{i-1} \frac{\partial \alpha_{i-1}}{\partial x_j} \right| L_{xi} \right) \| (z_1, z_2, \ldots, z_i) \|$$
$$\triangleq L_{\alpha_i} \, \| (z_1, z_2, \ldots, z_i) \| \tag{8.47}$$

where L_{α_i} depends on $(c_1, \ldots, c_i)$, and L_{x_i} depends on $(c_1, \ldots, c_{i-1})$. Then we have

$$\| (x_1, \ldots, x_n) \| = \left(\sum_{j=1}^{n} x_j^2 \right)^{1/2}$$
$$\leq \left(\sum_{j=1}^{n} L_{xj}^2 \, \| (z_1, z_2, \ldots, z_j) \|^2 \right)^{1/2}$$
$$\leq \left(\sum_{j=1}^{n} L_{xj}^2 \right)^{1/2} \| (z_1, \ldots, z_j) \|$$
$$\triangleq L_x \, \| (z_1, z_2, \ldots, z_n) \| \tag{8.48}$$

Remark 8.6   The properties (8.28)–(8.31) in Lemma 8.1 and 8.41 in Lemma 8.2 are key steps in the stability analysis, which will be used to eliminate the effects from state quantization.

The main results are formally stated in the following theorem.

Theorem 8.1
Consider the closed-loop adaptive system consisting of plant (8.1) with state quantization satisfying the bounded property (8.5), the adaptive backstepping controller (8.21) with parameter estimator with updating law (8.22), the following results can be guaranteed.

1. *All the closed-loop signals are globally uniformly bounded.*

2. *The upper bound of stabilization error $\| z(t) \|_{[0,T]}^2$ satisfies*

$$\| z(t) \|_{[0,T]}^2 = \frac{1}{T} \int_0^T \| z(t) \|^2 \, dt \leq \frac{2}{c} \left[\frac{V(0)}{T} + M \right] \tag{8.49}$$

if $k_\theta > \frac{2}{c} B^2$, where

$$c = \min\{c_1, c_2, \ldots, c_{n-1}, \frac{1}{4} c_n\} \tag{8.50}$$

$$M = \frac{1}{2} k_\theta \| (\theta - \theta_0) \|^2 + \frac{1}{c_n} \Delta_{\alpha_n}^2 + \frac{1}{c_n} \Delta_\psi^2$$

$$+ \frac{1}{c_n} \| \theta \|^2 \Delta_\phi^2 \tag{8.51}$$

$$V(0) = \sum_{j=1}^n \frac{1}{2} z_j^2(0) + \frac{1}{2} \tilde{\theta}(0)^T \Gamma^{-1} \tilde{\theta}(0) \tag{8.52}$$

$$B = L_\phi (L_x + \sqrt{n}\delta) \Delta_{zn}. \tag{8.53}$$

Proof 8.3   Considering the Lyapunov function

$$V = \sum_{j=1}^n \frac{1}{2} z_j^2 + \frac{1}{2} \tilde{\theta}^T \Gamma^{-1} \tilde{\theta} \tag{8.54}$$

then its derivative obtained by following the control design in (8.21)–(8.26) is given as

$$
\dot{V} = -\sum_{j=1}^{n-1} c_j z_j^2 + z_{n-1} z_n - \tilde{\theta}^T \Gamma^{-1} \dot{\hat{\theta}}
$$

$$
+ z_n \left(u(t) - \alpha_n + \alpha_n + \psi + \phi^T \theta - \dot{\alpha}_{n-1} \right)
$$

$$
= -\sum_{j=1}^{n-1} c_j z_j^2 - \tilde{\theta}^T \Gamma^{-1} \dot{\hat{\theta}} + z_n \left(\bar{\alpha}_n - \alpha_n + \alpha_n - \bar{\psi} - \hat{\theta}^T \bar{\phi} \right.
$$

$$
\left. + \psi + \theta^T \phi - \dot{\alpha}_{n-1} + z_{n-1} \right)
$$

$$
= -\sum_{j=1}^{n-1} c_j z_j^2 + z_n \left(\alpha_n - \dot{\alpha}_{n-1} + z_{n-1} \right) + z_n \left(\bar{\alpha}_n - \alpha_n \right)
$$

$$
+ z_n \left(\psi - \bar{\psi} \right) + z_n \left(\theta^T \phi - \hat{\theta} \bar{\phi} \right) - \tilde{\theta}^T \Gamma^{-1} \dot{\hat{\theta}}
$$

$$
\leq -\sum_{j=1}^{n} c_j z_j^2 - \frac{1}{2} k_\theta \parallel \tilde{\theta} \parallel^2 + \frac{1}{2} k_\theta \parallel (\theta - \theta_0) \parallel^2 + z_n \left(\bar{\alpha}_n - \alpha_n \right)
$$

$$
+ z_n \left(\psi - \bar{\psi} \right) + \left(\theta^T \phi z_n - \hat{\theta}^T \bar{\phi} z_n - \tilde{\theta}^T \bar{\phi} \bar{z}_n \right) \tag{8.55}
$$

where the property (8.27), is used. Using the properties (8.4), (8.29), (8.30), and (8.41), the last term in (8.55) satisfies the following inequality

$$
\theta^T \phi z_n - \hat{\theta}^T \bar{\phi} z_n - \tilde{\theta}^T \bar{\phi} \bar{z}_n
$$

$$
= \theta^T \phi z_n - \theta \bar{\phi} z_n + \tilde{\theta} \bar{\phi} z_n - \tilde{\theta} \bar{\phi} \bar{z}_n
$$

$$
\leq \parallel \theta \parallel |z_n| \Delta_\phi + \parallel \tilde{\theta} \parallel \parallel \bar{\phi} \parallel \Delta_{z_n}
$$

$$
\leq |z_n| \parallel \theta \parallel \Delta_\phi + \parallel \tilde{\theta} \parallel L_\phi \parallel (q(x), \dots, q(x_n)) \parallel \Delta_{z_n}
$$

$$
\leq |z_n| \parallel \theta \parallel \Delta_\phi + B \parallel \tilde{\theta} \parallel \parallel z \parallel \tag{8.56}
$$

where $z(t) = [z_1, z_2, \dots, z_n]^T$ and $B = L_\phi (L_x + \sqrt{n} \delta) \Delta_{z_n}$. Using the properties (8.28) and (8.31) in Lemma 8.1 and 8.56, the derivative of V is obtained as

$$
\dot{V} \leq -\sum_{j=1}^{n} c_j z_j^2 - \frac{1}{2} k_\theta \parallel \tilde{\theta} \parallel^2 + |z_n| \Delta_{\alpha_n} + |z_n| \Delta_\psi + |z_n| \parallel \theta \parallel \Delta_\phi
$$

$$
+ B \parallel \tilde{\theta} \parallel \parallel z \parallel + \frac{1}{2} k_\theta \parallel (\theta - \theta_0) \parallel^2
$$

$$
\leq -\sum_{j=1}^{n} c_j z_j^2 + \frac{3}{4} c_n z_n^2 + \frac{c}{2} \parallel z(t) \parallel^2 - \frac{1}{2} k_\theta \parallel \tilde{\theta} \parallel^2 + \frac{1}{2c} B^2 \parallel \tilde{\theta} \parallel^2 + M
$$

$$
\leq -\frac{c}{2} \parallel z(t) \parallel^2 - (\frac{1}{2} k_\theta - \frac{1}{2c} B^2) \parallel \tilde{\theta} \parallel^2 + M \tag{8.57}
$$

where c and M are defined in (8.50) and (8.51) and the Young's inequality was used as follows.

$$|z_n|\Delta_{\alpha_n} + |z_n|\Delta_\psi + |z_n| \parallel \theta \parallel \Delta_\phi$$

$$\leq \frac{3}{4}c_n|z_n|^2 + \frac{1}{c_n}\Delta_{\alpha_n}^2 + \frac{1}{c_n}\Delta_\psi^2 + \frac{1}{c_n} \parallel \theta \parallel^2 \Delta_\phi^2 \quad (8.58)$$

$$B \parallel \tilde{\theta} \parallel \parallel z \parallel \;\leq\; \frac{c}{2} \parallel z(t) \parallel^2 + \frac{1}{2c}B^2 \parallel \tilde{\theta} \parallel^2 \quad (8.59)$$

Choosing

$$k_\theta > \frac{2}{c}B^2 = \frac{2}{c}L_\phi^2(L_x + \sqrt{n}\delta)^2\Delta_{zn}^2, \quad (8.60)$$

(8.57) shows that

$$\dot{V} \leq -\frac{c}{2} \parallel z(t) \parallel^2 - \frac{1}{4}k_\theta \parallel \tilde{\theta} \parallel^2 + M \leq -\sigma V + M \quad (8.61)$$

where

$$\sigma \;=\; \min\{c, \frac{\frac{1}{2}k_\theta}{\lambda_{max}\left(\Gamma^{-1}\right)}\} \quad (8.62)$$

By direct integration of the above inequality, we have

$$V(t) \leq V(0)e^{-\sigma t} + \frac{M}{\sigma}(1 - e^{-\sigma t}) \leq V(0) + \frac{M}{\sigma} \quad (8.63)$$

which shows that V is uniformly bounded. Thus the signals $z_i(t)$ and $\hat{\theta}$ are bounded. From (8.8), (8.10), and (8.21), it further implies that $x_i(t)$ and $u(t)$ are bounded. Therefore all the closed-loop signals are globally uniformly bounded.

From (8.57), we have

$$\dot{V} \;\leq\; -\frac{c}{2} \parallel z(t) \parallel^2 + M \quad (8.64)$$

Integrating both sides of (8.64) yields that

$$\begin{aligned} \parallel z(t) \parallel_{[0,T]}^2 \;&=\; \frac{1}{T}\int_0^T \parallel z(t) \parallel^2 dt \\ &\leq\; \frac{2}{c}\left[\frac{V(0) - V(T)}{T} + M\right] \\ &\leq\; \frac{2}{c}\left[\frac{V(0)}{T} + M\right] \quad (8.65) \end{aligned}$$

From (8.50), (8.51), and (8.52), it follows that the upper bound of the overall stabilization errors in the mean square sense of (8.65) can be tuned by choosing suitable parameters k_θ, c_n, and Γ. Similarly, the bound of the parameter estimation error is obtained as

$$\parallel \tilde{\theta}(t) \parallel_{[0,T]}^2 = \frac{1}{T}\int_0^T \parallel \tilde{\theta}(t) \parallel^2 dt \leq \frac{4}{k_\theta}\left[\frac{V(0)}{T} + M\right] \quad (8.66)$$

After establishing the main results, we now highlight the main challenges in solving the problems caused by state quantization and key techniques proposed to handle them in the following remarks.

Remark 8.7

■ One major difficulty in dealing with state quantization is that the backstepping technique requires differentiating virtual controls and in turn the states by applying the chain rule. If the states are quantized, they become discontinuous and therefore it is difficult to analyze the resulting control system with the current backstepping-based approaches.

■ The above difficulty is overcome by not taking the derivative of $\bar{\alpha}_i$ in the controller design and stability analysis. Instead, the final control u in (8.21) and the virtual control law $\bar{\alpha}_i$ in (8.26) use the partial derivative term $\sum_{k=1}^{i-1} \frac{\partial \alpha_{i-1}}{\partial x_k} q(x_{k+1})$, which avoids taking the derivative of $\bar{\alpha}_{i-1}$ in the control design.

Remark 8.8
By following the general framework of backstepping procedure and including an additional term in the form of $-\Gamma k_\theta (\hat{\theta} - \theta_0)$ in the parameter adaptive law (8.22), we manage to design the backstepping-based adaptive control law.

Remark 8.9
The main challenge in stability analysis is how to handle the effects caused by analyzing states $x_1, \ldots, x_n$, while only the quantized states $q(x_1), \ldots, q(x_n(t))$ are used in the designed controller. More specifically, a major difficulty in stability analysis is how to compensate for the effects from the terms $z_n(\bar{\alpha}_n - \alpha_n)$, $z_n(\psi - \bar{\psi})$ and $(\theta^T \phi z_n - \hat{\theta}^T \bar{\phi} z_n - \tilde{\theta}^T \bar{\phi} \bar{z}_n)$ in (8.55). By establishing the properties (8.28)-(8.31) in Lemma 8.1, (8.41) in Lemma 8.2 and 8.56, such terms are bounded by functions depending only on the state z_i and parameter estimation error $\tilde{\theta}$. Thus all these effects can be compensated by two negative terms $-\sum_{j=1}^{n} c_j z_j^2$ and $-\frac{1}{2} k_\theta \parallel \tilde{\theta} \parallel^2$ as shown in (8.57). Above new techniques enable us to successfully overcome the difficulties caused by the discontinuity of quantized states and the bounded uncertainties resulted from the state quantization, so as to establish the results in Theorem 8.1.

Remark 8.10
As stated in Theorem 8.1, k_θ is chosen to satisfy (8.60), which depends on the quantization bound δ, the control parameters c_i, and L_ϕ. The lower bound of k_θ can be calculated with δ being known and therefore the designed adaptive controller is implementable. For simplicity, we let $\Gamma = \gamma I$. The upper bound of the overall stabilization errors in the mean square sense of (8.49) can be decreased by increasing γ and c_n.

Remark 8.11   The obtained bounds in Lemma 8.1 and Theorem 8.1 depend on the quantization bound δ. To reduce the conservatism of the results, we can design a quantizer by choosing a small quantization density.

8.4   Simulation Study

In this section we consider a pendulum system from [51] as shown in Figure 8.1. The equation of the motion for the pendulum system is represented as

$$ml\ddot{\theta} + mgsin(\theta) + kl\dot{\theta} = u(t) \tag{8.67}$$

where θ denotes the angle of the pendulum, m, l, and g are the mass $[kg]$, the length of the robe $[m]$, and the acceleration due to the gravity, k is an unknown friction coefficient, u represents an input torque provided by a DC motor. The states θ and $\dot{\theta}$ are quantized by a quantizer satisfying the bounding property (8.5). The objective is to design a control input for u to make the output θ track a reference signal $\theta_r(t) = sin(t)$. In the simulation, we consider three quantizers: uniform

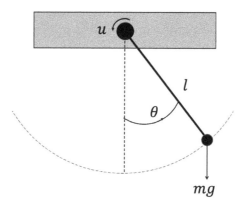

Figure 8.1 Pendulum

quantizer in (3.1), hysteresis-uniform quantizer in (3.3) and logarithmic-uniform quantizer in (3.5). The quantization parameters are chosen as $l = 0.1$ for uniform quantizer, $l = 0.1$ and $h = 0.5$ for hysteresis-uniform quantizer, and $l = 0.1$, $\rho = 0.05$ and $x_{th} = 0.8$ for logarithmic-uniform quantizer, respectively. We choose $x_1 = \theta - \theta_r$ and $x_2 = \dot{x}_1 = \dot{\theta} - \dot{\theta}_r$. The adaptive control law (8.21) and the parameter estimation (8.22) are used where $\frac{\partial \alpha_1}{\partial x_1} = -c_1$. The initial states are chosen as $x(0) = 0.2, \dot{x}(0) = 1$, and $\hat{\theta}(0) = 0.8$. The parameters in system (8.67) are selected as $m = 1, l = 1, g = 9.8$, and $k = 1$ for simulation. The design parameters are chosen as $c_3 = c_2 = 3, \gamma = 1$, and $k_\theta = 0.1$.

The trajectories of states θ and $\dot{\theta}$ and the control input are shown in Figures 8.2–8.3 for a uniform quantizer, Figures 8.4–8.5 for a hysteresis-uniform quantizer, and Figures 8.6–8.7 for a logarithmic-uniform quantizer, respectively. Clearly, the simulation results verify our theoretical findings in Theorem 8.1 and show the effectiveness of the proposed control scheme.

In addition, the range of outputs of the quantizer can also be calculated. For example, for the uniform quantizer with length $l = 0.1$, the state θ is bounded in $[-1\ 1]$ radian and the number of the output levels of the uniform quantizer is $round\left(\frac{\theta_{max} - \theta_{min}}{l}\right) = 20$.

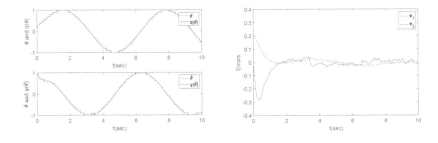

Figure 8.2 Uniform quantizer: θ and $\dot{\theta}$ and errors $\theta - \theta_r$ and $\dot{\theta} - \dot{\theta}_r$

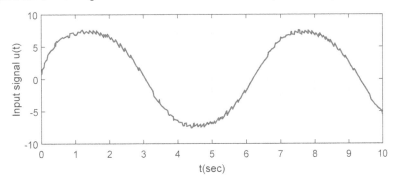

Figure 8.3 Uniform quantizer: Input $u(t)$

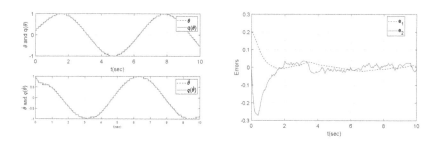

Figure 8.4 Uniform-Hysteresis quantizer: θ **and** $\dot{\theta}$ **and errors**

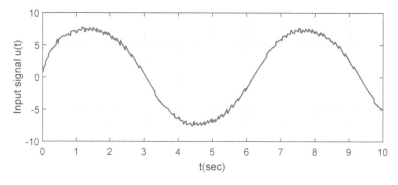

Figure 8.5 Uniform-Hysteresis quantizer: Input $u(t)$

8.5   Notes

In this chapter, we propose an adaptive feedback stabilization scheme for a class of nonlinear systems with state quantization. The nonlinear functions in the system satisfy the Lipschitz condition. The quantizers considered in this chapter satisfy the bounded property. By using backstepping approaches, a new adaptive control algorithm using only quantized states is developed through constructing a new compensation method for the effects of state quantization. By using a new approach for stability analysis, the global ultimate boundedness of the system is obtained. The stabilization error performance is also established and can be improved by appropriately adjusting design parameters.

Acknowledgment

Reprinted from Copyright (2021), with permission permission from IEEE. Jing Zhou, Changyun Wen, Wei Wang, and Fan Yang, "Adaptive Backstepping Control

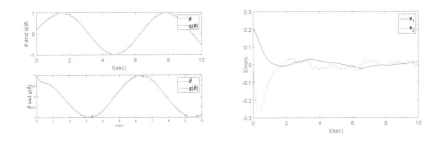

Figure 8.6 Logarithmic-Uniform quantizer: θ **and** $\dot{\theta}$ **and errors**

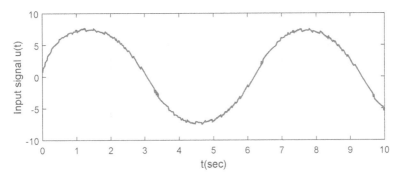

Figure 8.7 Logarithmic-Uniform quantizer: Input $u(t)$

of Nonlinear Uncertain Systems with Quantized States", *IEEE Transactions on Automatic Control*, vol. 64, no. 11, pp. 4756–4763, 2019.

Chapter 9

Adaptive Control of Systems with Sector-Bounded State Quantization

The result in Chapter 8 is only applicable to stabilization of uncertain nonlinear systems with bounded quantizers, where the quantization error of the state is bounded by a constant. By contrast, a sector-bounded quantizer as in Chapter 3 is considered in this chapter. Since the quantization error depends on the inputs of the state quantizer, they cannot be ensured bounded automatically. This constitutes the main challenge to handle the effects of state quantization in stability analysis. A new approach to stability analysis and handling discontinuity resulted from state quantization is developed by constructing a new compensation scheme for the effects of the sector-bounded state quantization. Besides, the projection technique is adopted to modify the design of adaptive law. Thus the closed-loop system stability can be achieved without the need for sufficient conditions dependent on the design parameters and quantization parameters.

9.1 Introduction

The stabilization problem for a class of uncertain high-order nonlinear systems with sector-bounded state quantization is investigated and a backstepping-based adaptive

DOI: 10.1201/9781003176626-9

control solution is provided. Although some techniques are also presented in Chapter 8 and [146] to handle the effects of state quantization. The main improvements of this chapter lie in the following two aspects.

- In Chapter 8, only the type of bounded quantizers are considered. By contrast, a more general type of sector-bounded state quantizers is considered in this chapter, such as logarithmic quantizer and hysteresis-logarithmic quantizer in Chapter 3. Since the quantization errors depend on the inputs of state quantizers, they cannot be ensured bounded automatically. This constitutes the main challenge to handle the effects of state quantization in stability analysis. By well establishing the relation between the input signal and error state, the closed-loop system stability can be achieved by choosing proper design parameters.

- The adaptive law is modified by adopting projection technique. Then for the case of bounded state quantizers, the sufficient condition ensuring the system stability, which relates to the adaptive gain and quantization parameters in Chapter 8, is successfully removed. This allows the adaptive gain and quantization parameters to be chosen as any arbitrary positive values.

9.2 Problem Statement

9.2.1 System Model

The same class of systems in Chapter 8 is under consideration.

$$x^{(n)}(t) = \psi\left(x, \dot{x}, \ldots, x^{(n-1)}\right) + \phi^T\left(x, \dot{x}, \ldots, x^{(n-1)}\right)\theta + u(t), \quad (9.1)$$

where $\bar{x}(t) \triangleq [x(t), \dot{x}(t), \ldots, x^{(n)}(t)]^T$ is the state of the system. $u(t) \in \Re^1$ is the control input. $\psi \in \Re^1$ and $\phi \in \Re^r$ are known nonlinear functions. $\theta \in \Re^r$ denotes the vector of constant parameters, which are assumed to be unknown. This class of uncertain nonlinear systems have been widely investigated in [46, 51, 71, 148].

In this chapter, only quantized states $(q(x), q(\dot{x}), \ldots, q(x^{(n-1)}))$ are measured, where $q(x)$ is a state quantizer whose the input is x. The feedback controller $u(t)$ in (9.1) only uses the quantized states, which is given by

$$u = u(q(x), q(\dot{x}), \ldots, q(x^{(n-1)})). \quad (9.2)$$

For system (9.1), the conditions of the existence of a unique solution are assumed to be satisfied. To generate control laws, we impose the following assumptions.

Assumption 9.1 *The functions ϕ and ψ satisfy the global Lipschitz continuity condition such that*

$$|\psi(y_1) - \psi(y_2)| \leq L_\psi \| y_1 - y_2 \| \quad (9.3)$$
$$\| \phi(y_1) - \phi(y_2) \| \leq L_\phi \| y_1 - y_2 \| \quad (9.4)$$

where L_ψ and L_ϕ *are positive constants,* $y_1, y_2 \in \Re^n$ *are real vectors. Note that Assumption 9.1 is also made in Chapter 8.*

Assumption 9.2 *The unknown parameter vector* θ *falls within a known compact convex set* C_θ *such that* $\parallel \theta \parallel \leq L_\theta$ *for any* $\theta \in C_\theta$ *and a positive constant* L_θ.

The control objective is to design an appropriate adaptive control law for $u(t)$ in system (9.1) by using only quantized states $\bar{x}^q = q(\bar{x}(t))$ such that all the closed-loop signals are uniformly bounded.

9.2.2   Sector-Bound Quantizer

The state quantizer is modeled as follows.

$$\bar{x}^q(t) \quad = \quad q(\bar{x}(t)) \tag{9.5}$$

In this chapter, the sector-bounded quantizer considered in the chapter has the following property, which is same as the property in (3.7).

$$|q(x) - x| \quad \leq \quad \delta \parallel x \parallel + \Delta \tag{9.6}$$

It is shown in Chapter 3 that logarithmic quantizer in (3.2) and hysteresis-logarithmic quantizer in (3.4) satisfy the sector-bounded property in (9.5).

Remark 9.1   Note that the bounded quantizer is considered in Chapter 8. By contrast, a more general type of sector-bounded quantizers is considered in this chapter. Since the quantization errors depend on the inputs of quantizers, they cannot be ensured bounded automatically. This constitutes the main challenge to handle the effects of state quantization in stability analysis.

9.3   Design of Adaptive Backstepping Controller

Define a group of new variables as $x_1 = x$, $x_i = x^{(i-1)}$, $i = 2, 3, \dots, n$. System (9.1) can be rewritten as

$$\begin{aligned}
\dot{x}_i &= x_{i+1}, \quad i = 1, \dots, n-1 \\
\dot{x}_n &= u(t) + \psi(x_1, \dots, x_n) + \theta^T \phi(x_1, \dots, x_n) \\
&= u(t) + \psi(\bar{x}) + \theta^T \phi(\bar{x})
\end{aligned} \tag{9.7}$$

where $\bar{x} = [x_1, x_2, \dots, x_n]^T$. $u(t)$ is the control input to be designed by utilizing only quantized states $\bar{x}^q$.

■   If the state $\bar{x}$ is not quantized, u and $\hat{\theta}$ are adopted to denote the final control input and parameter estimator introduced for the unknown vector θ. The adaptive controller can be designed by applying standard backstepping design technique in Section 8.3.1 in Chapter 8.

■ When state $x_i(t)$ is quantized with the quantizer $q(x_i)$, the quantized state is defined as

$$x_i^q = q(x_i), \; i = 1, 2, \ldots, n \tag{9.8}$$

Thus $\bar{x}^q(t)$ in (9.5) satisfies that $\bar{x}^q(t) = [x_1^q, \ldots, x_n^q]^T$. Define the error variables as

$$z_1^q = x_1^q \tag{9.9}$$
$$z_i^q = x_i^q - \alpha_{i-1}^q, \; i = 2, \ldots, n \tag{9.10}$$

where

$$\alpha_1^q = -c_1 z_1^q \tag{9.11}$$

$$\alpha_i^q = -c_i z_i^q - z_{i-1}^q + \sum_{k=1}^{i-1} \frac{\partial \alpha_{i-1}}{\partial x_k} x_{k+1}^q, i = 2, \ldots, n \tag{9.12}$$

where c_i are positive constants. The adaptive controller is designed as

$$u(t) = \alpha_n^q - \psi(\bar{x}^q) - \hat{\theta}^T \phi(\bar{x}^q)$$

$$= -c_n z_n^q - z_{n-1}^q + \sum_{k=1}^{n-1} \frac{\partial \alpha_{n-1}}{\partial x_k} x_{k+1}^q - \psi(\bar{x}^q) - \hat{\theta}^T \phi(\bar{x}^q) \tag{9.13}$$

$$\dot{\hat{\theta}} = Proj\{\Gamma \phi(\bar{x}^q) z_n^q\} \tag{9.14}$$

where $\hat{\theta}$ is the parameter estimator introduced for unknown vector θ, Γ is a positive definite matrix, $Proj\{.\}$ is the projector operator given in Appendix C. The partial derivatives $\frac{\partial \alpha_{n-i}}{\partial x_k}$, $i = 1, \ldots, n$, are calculated from functions α_i designed below if states are not quantized.

$$\alpha_1 = -c_1 z_1 \tag{9.15}$$

$$\alpha_i = -c_i z_i - z_{i-1} + \sum_{k=1}^{i-1} \frac{\partial \alpha_{i-1}}{\partial x_k} x_{k+1} \tag{9.16}$$

$$z_1 = x_1 \tag{9.17}$$
$$z_i = x_i - \alpha_{i-1}, \; i = 2, \ldots, n. \tag{9.18}$$

Remark 9.2 Note that only the quantized measurable states $x_i^q = q(x_i)$, $i = 1, \ldots, n$ can be utilized to generate the virtual control α_i^q in (9.12), final control signal u in (9.13) and the parameter updating law in (9.14). If we follow standard backstepping design procedure in Chapter 2, α_i^q in (9.12) will be designed in the form of $-c_i z_i^q - z_{i-1}^q + \dot{\alpha}_{i-1}^q$. However, since α_{i-1}^q involves quantized states x_i^q, α_{i-1}^q becomes discontinuous and its derivative cannot be computed. This obstacle is removed by not differentiating α_i^q in the process of adaptive control design. Instead,

α_i^q in (9.12) is constructed by utilizing the derivative terms $\sum\limits_{k=1}^{i-1} \frac{\partial \alpha_{i-1}}{\partial x_k} x_{k+1}^q$. As seen from (9.16), the partial derivatives $\frac{\partial \alpha_{i-1}}{\partial x_k}$ can be calculated because the virtual control α_i, designed for the case where the states are not quantized, is differentiable for x_k.

Remark 9.3   Note that the projector operator $Proj\{.\}$ is used in the parameter estimator (9.14) to ensure that $\| \hat{\theta} \| \leq L_\theta$. Let $\tilde{\theta} = \theta - \hat{\theta}$. It is also ensured that $\| \tilde{\theta} \| \leq L_\theta$. It is worth mentioning that the boundedness of θ, $\hat{\theta}$, and $\tilde{\theta}$ and the following property are helpful to guarantee the closed-loop system stability.

$$-\tilde{\theta}^T \Gamma^{-1} Proj\{\tau\} \leq -\tilde{\theta}^T \Gamma^{-1} \tau, \forall \hat{\theta} \in C_\theta, \ \theta \in C_\theta. \tag{9.19}$$

9.4   Stability Analysis

To analyze the closed-loop system stability, we first establish some preliminary results in the form of the following lemmas.

Lemma 9.1
The effects of state quantization are bounded by functions of z as follows:

$$|\psi(\bar{x}^q) - \psi(\bar{x})| \leq \Delta\Delta_\psi^1 + \delta\Delta_\psi^2 \| z \| \tag{9.20}$$

$$\| \phi(\bar{x}^q) - \phi(\bar{x}) \| \leq \Delta\Delta_\phi^1 + \delta\Delta_\phi^2 \| z \| \tag{9.21}$$

$$|z_i^q - z_i| \leq \Delta\Delta_{z_i}^1 + \delta\Delta_{z_i}^2 \| \bar{z}_i \|, \ i = 1, \dots, n \tag{9.22}$$

$$|\alpha_i^q - \alpha_i| \leq \Delta\Delta_{\alpha_i}^1 + \delta\Delta_{\alpha_i}^2 \| \bar{z}_i \|, \ i = 1, \dots, n \tag{9.23}$$

$$\| z^q - z \| \leq \Delta\Delta_z^1 + \delta\Delta_z^2 \| z \| \tag{9.24}$$

where $\bar{z}_i = [z_1, \dots, z_i]^T$, and

$$\Delta_\psi^1 = \Delta_\psi^2 = L_\psi \tag{9.25}$$

$$\Delta_\phi^1 = \Delta_\phi^2 = L_\phi \tag{9.26}$$

$$\Delta_{z_1}^1 = \Delta_{z_1}^2 = 1 \tag{9.27}$$

$$\Delta_{\alpha_1}^1 = \Delta_{\alpha_1}^2 = c_1 \tag{9.28}$$

$$\Delta_{z_i}^1 = 1 + \Delta_{\alpha_{i-1}}^1, \ i = 2, \dots, n \tag{9.29}$$

$$\Delta_{z_i}^2 = L_{x_i} + \Delta_{\alpha_{i-1}}^2, \ i = 2, \dots, n \tag{9.30}$$

$$\Delta_{\alpha_i}^1 = c_i \Delta_{z_i}^1 + \Delta_{z_{i-1}}^1 + \sum\limits_{k=1}^{i-1} \left| \frac{\partial \alpha_{i-1}}{\partial x_k} \right|, i = 2, \dots, n \tag{9.31}$$

$$\Delta^2_{\alpha_i} = c_i \Delta^2_{z_i} + \Delta^2_{z_{i-1}} + \sum_{k=1}^{i-1} \left| \frac{\partial \alpha_{i-1}}{\partial x_k} \right| L_{x_k}, i = 2, \ldots, n \qquad (9.32)$$

$$\Delta^j_z = \left(\sum_{i=1}^n (\Delta^j_{z_i})^2 \right)^{1/2}, j = 1, 2 \qquad (9.33)$$

Proof 9.1 Using Lipschitz conditions of ψ and ϕ in Assumption 9.2, the following expressions can be derived.

$$|\psi(\bar{x}^q) - \psi(\bar{x})| \le L_\psi \| \bar{x}^q - \bar{x} \|$$
$$\le L_\psi \Delta + L_\psi \delta \| \bar{x} \|$$
$$\le L_\psi \Delta + L_\psi \delta L_x \|z\| \triangleq \Delta\Delta^1_\psi + \delta\Delta^2_\psi \| z \| \qquad (9.34)$$
$$\| \phi(\bar{x}^q) - \phi(\bar{x}) \| \le L_\phi \| \bar{x}^q - \bar{x} \|$$
$$\le L_\phi \Delta + L_\phi \delta L_x \| z \| \triangleq \Delta\Delta^1_\phi + \delta\Delta^2_\phi \| z \| . \qquad (9.35)$$

From (9.9)–(9.12), it is shown that

$$|z^q_1 - z_1| = |x^q_1 - x_1| \le \Delta + \delta|x_1| \triangleq \Delta\Delta^1_{z_1} + \delta\Delta^2_{z_1} \| z_1 \| \qquad (9.36)$$
$$|\alpha^q_1 - \alpha_1| = |-c_1(z^q_1 - z_1)| \le c_1 \Delta\Delta^1_{z_1} + c_1 \delta\Delta^2_{z_1} \| z_1 \| \triangleq \Delta\Delta^1_{\alpha_1} + \delta\Delta^2_{\alpha_1} \| z_1 \| \qquad (9.37)$$

$$|z^q_2 - z_2| = |(x^q_2 - x_2) - (\alpha^q_1 - \alpha_1)|$$
$$\le \Delta + \delta L_{x_2} \|\bar{z}_2\| + \Delta\Delta^1_{\alpha_1} + \delta\Delta^2_{\alpha_1} \| z_1 \|$$
$$\le \Delta\Delta^1_{z_2} + \delta\Delta^2_{z_2} \| \bar{z}_2 \| \qquad (9.38)$$
$$|\alpha^q_2 - \alpha_2| = \left| -c_2(z^q_2 - z_2) - (z^q_1 - z_1) + \frac{\partial\alpha_1}{\partial x_1}(x^q_2 - x_2) \right|$$
$$\le \Delta \left(c_2 \Delta^1_{z_2} + \Delta^1_{z_1} + \left| \frac{\partial\alpha_1}{\partial x_1} \right| \right) + \delta \left(c_2 \Delta^2_{z_2} + \Delta^2_{z_1} + \left| \frac{\partial\alpha_1}{\partial x_1} \right| L_{x_2} \right) \| \bar{z}_2 \|$$
$$\triangleq \Delta\Delta^1_{\alpha_2} + \delta\Delta^2_{\alpha_2} \| \bar{z}_2 \| \qquad (9.39)$$

where $\bar{z}_i = [z_1, \ldots, z_i]^T$. Along the analysis lines of z_i in (9.18), α_i in (9.16), z^q_i in (9.10), α^q_i in (9.12), we have

$$|z^q_i(\bar{x}^q_i) - z_i(\bar{x}_i)| = |(x^q_i - x_i) - (\alpha^q_{i-1} - \alpha_{i-1})|$$
$$\le \Delta + \delta L_{x_i} \| \bar{z}_i \| + \Delta\Delta^1_{\alpha_{i-1}} + \delta\Delta^2_{\alpha_{i-1}} \| \bar{z}_{i-1} \|$$
$$\le \Delta\Delta^1_{z_i} + \delta\Delta^2_{z_i} \| \bar{z}_i \| \qquad (9.40)$$

and,

$$|\alpha_i^q(\bar{x}_i^q) - \alpha_i(\bar{x}_i^q)|$$

$$\leq \left| -c_i(z_i - z_i^q) - (z_{i-1}^q - z_{i-1}) \right| + \left| \sum_{k=1}^{i-1} \frac{\partial \alpha_{i-1}}{\partial x_k} \left(x_{k+1}^q - x_{k+1} \right) \right|$$

$$\leq \Delta \left(c_i \Delta_{z_i}^1 + \Delta_{z_{i-1}}^1 + \sum_{k=1}^{i-1} \left| \frac{\partial \alpha_{i-1}}{\partial x_k} \right| \right)$$

$$+ \delta \left(c_i \Delta_{z_i}^2 + \Delta_{z_{i-1}}^2 + \sum_{k=1}^{i-1} \left| \frac{\partial \alpha_{i-1}}{\partial x_k} \right| L_{x_k} \right) \parallel \bar{z}_i \parallel$$

$$\leq \Delta \Delta_{\alpha_i}^1 + \delta \Delta_{\alpha_i}^2 \parallel \bar{z}_i \parallel \tag{9.41}$$

$$\parallel z^q - z \parallel = \left(\sum_{i=1}^{n} |z_i^q - z_i|^2 \right)^{1/2}$$

$$\leq \left(\sum_{i=1}^{n} (\Delta \Delta_{z_i}^1)^2 \right)^{1/2} + \left(\sum_{i=1}^{n} (\delta \Delta_{z_i}^2)^2 \right)^{1/2} \parallel z \parallel$$

$$\triangleq \Delta \Delta_z^1 + \delta \Delta_z^2 \parallel z \parallel \tag{9.42}$$

Lemma 9.2
The state $\bar{x} = [x_1, \ldots, x_n]^T$, $\bar{x}^q = [x_1^q, \ldots, x_n^q]^T$, and the control $u(t)$ in (9.13) satisfy the following inequalities:

$$\parallel \bar{x} \parallel \quad \leq \quad L_x \parallel z \parallel \tag{9.43}$$
$$\parallel \bar{x}^q \parallel \quad \leq \quad (1 + \delta) L_x \parallel z \parallel + \Delta \tag{9.44}$$
$$|u| \quad \leq \quad L_u \parallel z^q \parallel \tag{9.45}$$

where $z = [z_1, \ldots, z_n]^T$, $z^q = [z_1^q, \ldots, z_n^q]$. L_x and L_u are positive constants defined as follows, which relate to the design parameters $c_1, \ldots, c_n$.

$$L_x \quad = \quad \left(\sum_{i=1}^{n} L_{x_i}^2 \right)^{\frac{1}{2}} \tag{9.46}$$

$$L_u \quad = \quad L_{\alpha_n} + L_\psi L_x + L_\theta L_\phi L_x, \tag{9.47}$$

where

$$L_{\alpha_1} \quad = \quad c_1 \tag{9.48}$$
$$L_{x_i} \quad = \quad 1 + L_{\alpha_{i-1}}, \quad i = 2, \ldots, n \tag{9.49}$$

$$L_{\alpha_i} \quad = \quad c_i + 1 + \sum_{k=1}^{i-1} \left| \frac{\partial \alpha_{i-1}}{\partial x_k} \right| L_{x_i}, i = 2, \ldots, n. \tag{9.50}$$

Proof 9.2 From the definitions α_i in (9.15)–(9.16) and z_i in (9.17)–(9.18), it can be derived that

$$|x_1| = |z_1| \tag{9.51}$$

$$|\alpha_1| \le c_1|z_1| \triangleq L_{\alpha_1}|z_1| \tag{9.52}$$

$$|x_2| \le |z_2 + \alpha_1| \le |z_2| + L_{\alpha_1}|z_1|$$
$$\le (1 + L_{\alpha_1}) \| [z_1, z_2]^T \| \triangleq L_{x2} \| [z_1, z_2]^T \| \tag{9.53}$$

$$|\alpha_2| \le c_2|z_2| + |z_1| + \left| \frac{\partial \alpha_1}{\partial x_1} x_2 \right|$$
$$\le \left(c_2 + 1 + \left| \frac{\partial \alpha_1}{\partial x_1} \right| L_{x_2} \right) \| [z_1, z_2]^T \| \triangleq L_{\alpha_2} \| [z_1, z_2] \| \tag{9.54}$$

Following the similar procedure for $i = 1, 2, \ldots, n-1$, we have

$$|x_i| \le |z_i + \alpha_{i-1}| \le |z_i| + L_{\alpha_{i-1}} \| [z_1, \ldots, z_{i-1}]^T \|$$
$$\le (1 + L_{\alpha_{i-1}}) \| [z_1, z_2, \ldots, z_i]^T \| \triangleq L_{x_i} \| [z_1, z_2, \ldots, z_i]^T \| \tag{9.55}$$

$$|\alpha_i| \le c_i|z_i| + |z_{i-1}| + \left| \sum_{k=1}^{i-1} \frac{\partial \alpha_{i-1}}{\partial x_k} x_{k+1} \right|$$
$$\le \left(c_i + 1 + \sum_{k=1}^{i-1} \left| \frac{\partial \alpha_{i-1}}{\partial x_k} \right| L_{x_i} \right) \| [z_1, z_2, \ldots, z_i]^T \|$$
$$\triangleq L_{\alpha_i} \| [z_1, z_2, \ldots, z_i]^T \| \tag{9.56}$$

$$\| \bar{x} \| = \left(\sum_{i=1}^{n} x_i^2 \right)^{1/2} \le \left(\sum_{i=1}^{n} L_{xi}^2 \| [z_1, z_2, \ldots, z_i]^T \|^2 \right)^{1/2}$$
$$\le \left(\sum_{i=1}^{n} L_{xi}^2 \right)^{\frac{1}{2}} \| z(t) \| \triangleq L_x \| z(t) \| \tag{9.57}$$

In view of (9.6), $\bar{x}^q$ satisfies

$$\| \bar{x}^q \| \le \| \bar{x} \| + \delta \| \bar{x} \| + \Delta \le (1 + \delta) L_x \| z \| + \Delta \tag{9.58}$$

From (9.9)–(9.12), by following the analysis in (9.51)–(9.57), we obtain that

$$|x_i^q| \le L_{x_i} \| [z_1^q, z_2^q, \ldots, z_i^q]^T \| \tag{9.59}$$

$$|\alpha_i^q| \le L_{\alpha_i} \| [z_1^q, z_2^q, \ldots, z_i^q]^T \| \tag{9.60}$$

$$\| \bar{x}^q \| \le L_x \| z^q \| \tag{9.61}$$

From (9.13) and (9.59)–(9.61), we have

$$|u| \le |\alpha_n^q| + |\psi(\bar{x}^q)| + \| \hat{\theta} \| |\phi(\bar{x}^q)|$$
$$\le L_{\alpha_n} \| z^q \| + L_\psi \| \bar{x}^q \| + L_\theta L_\phi \| \bar{x}^q \|$$
$$\le (L_{\alpha_n} + L_\psi L_x + L_\theta L_\phi L_x) \| z^q \| \triangleq L_u \| z^q \| \tag{9.62}$$

Based on Lemma 9.2 and Lemma 9.1, the main results are formally stated in the following theorem.

Theorem 9.1

Consider the closed-loop system consisting of system (9.1) with state quantization (9.5) satisfying the property (9.6), the designed adaptive controller (9.13) and parameter estimator update law (9.14). If the parameters satisfy

$$c - \delta\beta - 3r \geq \epsilon > 0, \tag{9.63}$$

where ϵ and r are positive constants, and

$$c = \min\{c_1, c_2, \dots, c_{n-1}, c_n\} \tag{9.64}$$

$$\beta = \Delta_{\alpha_n}^2 + \Delta_\psi^2 + B_2, \tag{9.65}$$

The following results can be guaranteed.

■ *All the closed-loop signals are uniformly bounded.*

■ *The stabilization error $\| z(t) \|$ is ultimately bounded as follows*

$$\| z(t) \| \leq \sqrt{\frac{M}{\epsilon}} \tag{9.66}$$

where

$$M = (\Delta)^2 \left(\frac{(\Delta_{\alpha_n}^1)^2}{4r} + \frac{(\Delta_\psi^1)^2}{4r} + \frac{(B_1)^2}{4r} + B_0 \right) \tag{9.67}$$

with

$$B_0 = L_\theta L_\phi \Delta_{z_n}^1 \tag{9.68}$$

$$B_1 = L_\theta \Delta_\phi^1 + L_\theta L_\phi \delta \Delta_{z_n}^2 + L_\theta L_\phi (1 + \delta) L_x \Delta_{z_n}^1 \tag{9.69}$$

$$B_2 = L_\theta \Delta_\phi^2 + L_\theta L_\phi (1 + \delta) L_x \Delta_{z_n}^2. \tag{9.70}$$

Proof 9.3 The Lyapunov function for the entire closed-loop system is defined as

$$V = \sum_{i=1}^n \frac{1}{2} z_i^2 + \frac{1}{2} \tilde{\theta}^T \Gamma^{-1} \tilde{\theta}, \tag{9.71}$$

where z_i is given in (9.17)–(9.18) and $\tilde{\theta} = \theta - \hat{\theta}$. From (9.13)–(9.14), the derivative of V is calculated as

$$
\begin{aligned}
\dot{V} &= -\sum_{i=1}^{n-1} c_i z_i^2 + z_{n-1}z_n - \tilde{\theta}^T \Gamma^{-1}\dot{\hat{\theta}} + z_n\left(u(t) + \psi + \phi^T\theta - \dot{\alpha}_{n-1}\right) \\
&= -\sum_{i=1}^{n-1} c_i z_i^2 - \tilde{\theta}^T\Gamma^{-1}\dot{\hat{\theta}} + z_n\left(\alpha_n^q - \psi(\bar{x}^q) - \hat{\theta}^T\phi(\bar{x}^q)\right. \\
&\qquad \left. -\alpha_n + \alpha_n + \psi + \theta^T\phi - \dot{\alpha}_{n-1} + z_{n-1}\right) \\
&= -\sum_{i=1}^{n-1} c_i z_i^2 + z_n\left(\alpha_n - \dot{\alpha}_{n-1} + z_{n-1}\right) + z_n\left(\alpha_n^q - \alpha_n\right) \\
&\quad +z_n\left(\psi(\bar{x}) - \psi(\bar{x}^q)\right) + z_n\left(\theta^T\phi(\bar{x}) - \hat{\theta}\phi(\bar{x}^q)\right) - \tilde{\theta}^T\Gamma^{-1}\dot{\hat{\theta}} \\
&\leq -\sum_{i=1}^{n} c_i z_i^2 + |z_n||\alpha_n^q - \alpha_n| + |z_n||\psi(\bar{x}) - \psi(\bar{x}^q)| \\
&\quad +\left(\theta^T\phi(\bar{x})z_n - \hat{\theta}^T\phi(\bar{x}^q)z_n - \tilde{\theta}^T\phi(\bar{x}^q)z_n^q\right)
\end{aligned}
\tag{9.72}
$$

Using the properties (9.4), (9.6), (9.44) in Lemma 9.2, and (9.21) and (9.22) in Lemma 9.1, the last three terms in (9.72) satisfies the following inequality

$$
\begin{aligned}
&\theta^T\phi(\bar{x})z_n - \hat{\theta}^T\phi(\bar{x}^q)z_n - \tilde{\theta}^T\phi(\bar{x}^q)z_n^q \\
&= \theta^T\phi(\bar{x})z_n - \theta^T\phi(\bar{x}^q)z_n + \tilde{\theta}\phi(\bar{x}^q)z_n - \tilde{\theta}\phi(\bar{x}^q)z_n^q \\
&\leq \|\theta\|\,|z_n||\phi(\bar{x}) - \phi(\bar{x}^q)| + \|\tilde{\theta}\|\,\|\phi(\bar{x}^q)\|\,|z_n - z_n^q| \\
&\leq L_\theta\|z\|(\Delta\Delta_\phi^1 + \delta\Delta_\phi^2\|z\|) + L_\theta L_\phi\|\bar{x}^q\|(\Delta\Delta_{z_n}^1 + \delta\Delta_{z_n}^2\|z\|) \\
&\leq L_\theta L_\phi\Delta_{z_n}^1(\Delta)^2 + \Delta\left(L_\theta\Delta_\phi^1 + L_\theta L_\phi(1+\delta)L_x\Delta_{z_n}^1 + L_\theta L_\phi\delta\Delta_{z_n}^2\right)\|z\| \\
&\quad +\delta\left(L_\theta\Delta_\phi^2 + L_\theta L_\phi(1+\delta)L_x\Delta_{z_n}^2\right)\|z\|^2 \\
&\leq (\Delta)^2 B_0 + \Delta B_1\|z\| + \delta B_2\|z\|^2
\end{aligned}
\tag{9.73}
$$

where $B_j, j = 0, 1, 2$, are defined in (9.68)–(9.70).
Using the properties (9.20), (9.23), (9.73) and the Young's inequality with positive parameter r (i.e. $|ab| \leq ra^2 + \frac{b^2}{4r}$), (9.72) is further computed as

$$
\begin{aligned}
\dot{V} &\leq -\sum_{i=1}^{n} c_i z_i^2 + \Delta\Delta_{\alpha_n}^1|z_n| + \delta\Delta_{\alpha_n}^2\|z\|^2 \\
&\quad +\Delta\Delta_\psi^1|z_n| + \delta\Delta_\psi^2\|z\|^2 + B_0(\Delta)^2 + \Delta B_1\|z\| + \delta B_2\|z\|^2 \\
&\leq -\left(c - \delta(\Delta_{\alpha_n}^2 + \Delta_\psi^2 + B_2) - 3r\right)\|z(t)\|^2 + M \\
&= -(c - \delta\beta - 3r)\|z(t)\|^2 + M \\
&\leq -\epsilon\|z(t)\|^2 + M
\end{aligned}
\tag{9.74}
$$

where the inequality $c - \delta\beta - 3r \geq \epsilon > 0$ in (9.63) has been used. ϵ, $r_i(i =$

$1, \ldots, 5$) are positive constants and c, β, and M are defined in (9.64), (9.65), and (9.67), respectively. It is shown from (9.74) that $\dot{V} < 0, \forall \parallel z(t) \parallel > \sqrt{\frac{M}{\epsilon}}$. Thus the ultimate bound of $z(t)$ satisfies (9.66).

From (9.24) and the boundedness of z, z^q is bounded. Thus x_1^q and α_1^q in (9.11) is bounded. From (9.10), x_2^q is bounded. Thus α_2^q is bounded. By the same token, the boundedness of x^q and α_i^q for $i = 1, \ldots, n$ can be shown. The boundedness of $\hat{\theta}$ is ensured by the projection operator (9.14) as discussed in Remark 9.3. Based on Assumption 9.1, it implies that that $u(t)$ in (9.13) is bounded. Therefore, the boundedness of all the closed-loop signals can be ensured.

Remark 9.4   Note that for the given design parameters c_i, the choice of quantization parameters δ and Δ is arbitrary, provided that the inequality (9.63) holds. Therefore, (9.63) provides some insights on how to choose these quantization parameters. Besides, Theorem 9.1 implies that the number of quantization levels for δ_i is finite since all the closed-loop signals are bounded. It is noted that β is computable from the definition (9.65), which depends on the design parameters c_i, the system parameters in Lemma 9.2 and Lemma 9.1 , and L_ψ, L_ϕ which are assumed to be known in Assumption 9.1.

Remark 9.5   From (9.67) and (9.68)–(9.69), the upper bound of the stabilization error in the sense of (9.66) can be decreased if the quantization parameters δ and Δ are decreased while all design parameters c_i are kept unchanged.

9.5   Simulation Results

We consider the same pendulum system with state quantization as in Chapter 8 as follows.

$$ml\ddot{\vartheta} + mg\sin(\vartheta) + kl\dot{\vartheta} \;=\; bu(t) \tag{9.75}$$
$$\bar{\vartheta}^q \;=\; q_{log}(\vartheta; \dot{\vartheta}) \tag{9.76}$$

where $q_{log}(\vartheta; \dot{\vartheta})$ is a logarithmic quantizer in (3.2), ϑ is the angle, m and l are the mass $[kg]$ and length of the robe $[m]$, g is the gravity acceleration, k denotes an unknown friction coefficient, $u(t)$ denotes an input torque, b denotes the control gain. The system parameters are chosen as $m = 3kg$, $l = 5m$, and $g = 9.8m/s^2$. The friction coefficient is set as $k = 3$. The control gain is set as $b = 15$. Define $\theta = \frac{k}{m}$ which is an unknown system parameter.

The objective is to design u by using only the quantized states $\bar{\vartheta}^q$ such that the closed-loop system is stable. The adaptive controller (9.13) and parameter update law (9.14) are adopted. In the simulation, the initials are set as $\vartheta(0) = \dot{\vartheta}(0) = 0.8$ and $\dot{\theta}(0) = 0$. In the simulation, the states ϑ, $\dot{\vartheta}$ are quantized by the logarithmic quantizer with quantization parameters $\delta = 0.07$ and $y_{min} = 0.1$. The design parameters are chosen as $c_1 = c_2 = 2$ and $\Gamma = 1$. The performance of ϑ, $\dot{\vartheta}$ and the quantized states $q_{log}(\vartheta)$, $q_{log}(\dot{\vartheta})$ with respect to the time is shown in Figure 9.1. Figure 9.2 shows the designed input $v(t)$. Clearly, all the states and input are bounded.

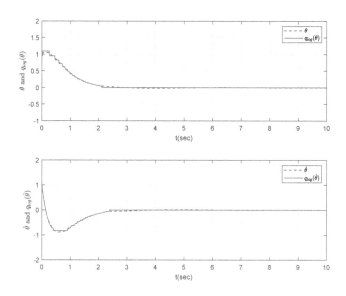

Figure 9.1 The states ϑ, $\dot{\vartheta}$ and the quantized states $q(\vartheta)$, $q(\dot{\vartheta})$

9.6 Notes

In this chapter, adaptive backstepping controllers have been designed for uncertain nonlinear systems with sector-bounded state quantization. Compared to the result in Chapter 8, a more general type of sector-bounded state quantizers is considered. Besides, the projection technique is adopted to modify the design of the adaptive law. Thus the closed-loop system stability can be achieved without the need for a sufficient condition dependent on the design parameters and quantization parameters.

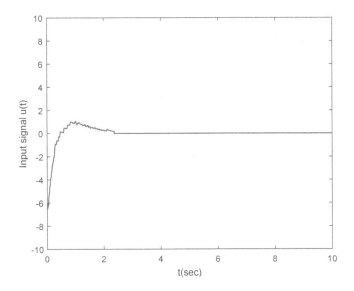

Figure 9.2 The designed input $u(t)$

INPUT AND STATE/OUTPUT QUANTIZATION COMPENSATION

Chapter 10

Adaptive State Feedback Control of Systems with Input and State Quantization

It is common in network control systems that the sensor and control signals are transmitted via a common communication network. In this chapter, an adaptive backstepping-based control algorithm is proposed for uncertain nonlinear systems with both input and state quantization, where the quantizers satisfy the sector-bounded property. In addition to overcoming the difficulties in proceeding recursive design of virtual controls with quantized states, the relation between the input control signal and error states is well established to handle the effects of both input and state quantization. The closed-loop system stability is achieved without the need for a sufficient condition dependent on the design parameters and quantization parameters. It is shown that all closed-loop signals are ensured uniformly bounded and all states will converge to a compact set.

10.1 Introduction

Quantized control has attracted considerable attention in recent years, due to its theoretical and practical importance in network control systems. A great number of representative results have been reported on analysis and control of quantized

DOI: 10.1201/9781003176626-10

feedback systems, as can be observed in [3, 33, 34, 59, 65, 99, 124, 145–147]. However, most of the results are concerned with either input quantization in Chapters 4–7 and in [33, 34, 59, 124, 145, 147], or state quantization in Chapters 8–9 and in [3, 65, 99, 146].

This chapter is concerned with the adaptive backstepping control problem for uncertain nonlinear systems with both input and state quantization. So far, only a few works have been reported to handle the issue with simultaneous existence of quantizers in both uplink and downlink communication channels of networked control systems. References [7, 8, 17, 82, 131] are some examples, however only linear systems are considered. Such a problem is important as it is common in general networked systems that the sensor and the actuator are connected via a shared communication network. In such quantized control systems, the state measurements and the control signals are both processed by quantizers.

The main features of this chapter are summarized as follows.

- The stabilization problem for a class of uncertain high-order nonlinear systems with both input and state quantization is investigated and a backstepping-based adaptive control solution is provided. Compared with the existing results investigating only input quantization, the main challenge is that only quantized states can be utilized to construct the virtual control signal in each recursive step. Hence the virtual controls are discontinuous, of which the derivatives cannot be computed as often done in standard backstepping design procedure. To overcome the difficulty, differentiable virtual controls are firstly designed by assuming the states are not quantized. Their partial derivatives multiplied by the quantized states are then utilized to complete the design of virtual controls for the case of state quantization.

- Note that some techniques are also presented in Chapter 9 to handle the effects of state quantization. However, only state quantization is considered in Chapter 9. Since the quantization errors depend on both input and state quantization, they cannot be ensured bounded automatically. This constitutes the main challenge to handle the effects of both state and input quantization in stability analysis. By well establishing the properties related to the input and output of the quantizers, the closed-loop system stability can be achieved by choosing proper design parameters.

10.2   Problem Statement

10.2.1   System Model

In this chapter, a quantized feedback system is considered, as shown in Figure 10.1. The system under consideration is represented by

$$x^{(n)}(t) = \psi\left(x, \dot{x}, \ldots, x^{(n-1)}\right) + \phi^T\left(x, \dot{x}, \ldots, x^{(n-1)}\right)\theta + u(t), \qquad (10.1)$$

where $x^{(i)}(t) \in \Re^1$, $i = 0, 1, \ldots, n$ are the states of the system. $u(t) \in \Re^1$ is the control input. $\psi \in \Re^1$ and $\phi \in \Re^r$ are known nonlinear functions. $\theta \in \Re^r$ denotes the vector of constant parameters, which are assumed to be unknown.

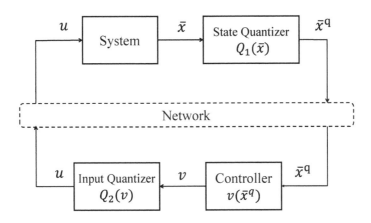

Figure 10.1 Input and state quantized control systems

The state $\bar{x}(t) \triangleq [x(t), \dot{x}(t), \ldots, x^{(n)}(t)]^T$ and the designed control signal $v(t)$ are quantized at the encoder side. The two quantizers are modeled as follows.

$$\bar{x}^q(t) = Q_1(\bar{x}(t)) \tag{10.2}$$
$$u(t) = Q_2(v(t)) \tag{10.3}$$

For system (10.1), the existence of a unique solution is assumed. To generate control laws, we impose the following assumptions.

Assumption 10.1 *The functions ϕ and ψ are globally Lipschitz continuous. That is,*

$$|\psi(y_1) - \psi(y_2)| \leq L_\psi \| y_1 - y_2 \| \tag{10.4}$$
$$\| \phi(y_1) - \phi(y_2) \| \leq L_\phi \| y_1 - y_2 \| \tag{10.5}$$

where L_ψ and L_ϕ are positive constants, $y_1, y_2 \in \Re^n$ are real vectors.

Assumption 10.2 *The unknown parameter vector θ falls within a known compact convex set C_θ such that $\| \theta \| \leq L_\theta$ for any $\theta \in C_\theta$ and a positive constant L_θ.*

The control objective is to design a quantized controller $u = Q_2(v)$ with an appropriate adaptive control law for $v(t)$ in system (10.1) by using only quantized states $\bar{x}^q = Q_1(x_1, x_2, \ldots, x_n)$ such that all the closed-loop signals are uniformly bounded.

10.2.2 Quantizer Model

In this chapter, sector-bounded quantizers are considered for both input and state quantization, which have the following property.

$$
\begin{aligned}
|Q_1(\bar{x}) - \bar{x}| &\leq \delta_1 \| \bar{x} \| + \Delta_1 \\
|Q_2(v) - v| &\leq \delta_2 \| v \| + \Delta_2,
\end{aligned}
\tag{10.6}
$$

where $0 < \delta_i < 1$ and $\Delta_i > 0$ for $i = 1, 2$ are quantization parameters. It is shown in Chapter 3 that logarithmic quantizer in (3.2) and hysteresis-logarithmic quantizer in (3.4) satisfy the sector-bounded property in (10.6).

10.3 Design of Adaptive Backstepping Controller

Define a group of new variables as $x_1 = x$, $x_i = x^{(i-1)}$, $i = 2, 3, \ldots, n$. System (10.1) can be rewritten as

$$
\begin{aligned}
\dot{x}_i &= x_{i+1}, \quad i = 1, \ldots, n-1 \\
\dot{x}_n &= u(t) + \psi(x_1, \ldots, x_n) + \theta^T \phi(x_1, \ldots, x_n) \\
&= u(t) + \psi(\bar{x}) + \theta^T \phi(\bar{x})
\end{aligned}
\tag{10.7}
$$

where $\bar{x} = [x_1, x_2, \ldots, x_n]^T$. $u(t)$ is the quantized input with $u(t) = Q_2(v(t))$, where $v(t)$ denotes the control input to be designed by utilizing only quantized states $\bar{x}^q$.

10.3.1 System without Quantization

If the states $\bar{x}$ and input v are not quantized, v_0 and $\hat{\theta}_0$ are adopted to denote the final control input and parameter estimator introduced for the unknown vector θ. The adaptive controller can be designed by applying standard backstepping design technique.

Introduce the change of coordinates as

$$
\begin{aligned}
z_1 &= x_1 \tag{10.8} \\
z_i &= x_i - \alpha_{i-1}, \ i = 2, \ldots, n. \tag{10.9}
\end{aligned}
$$

α_{i-1} is the virtual control function chosen as follows for each step i, which is a function of $x_1, \ldots, x_{i-1}$.

$$
\alpha_1 = -c_1 z_1
\tag{10.10}
$$

$$
\alpha_i = -c_i z_i - z_{i-1} + \sum_{k=1}^{i-1} \frac{\partial \alpha_{i-1}}{\partial x_k} x_{k+1}, \ i = 2, \ldots, n
\tag{10.11}
$$

where c_i, $1 \leq i \leq n$, are positive constant design parameters. As explained in [146], it can be easily shown that $\frac{\partial \alpha_{i-1}}{\partial x_k}$ are constants depending on $c_1, \ldots, c_{i-1}$. For examples, $\frac{\partial \alpha_1}{\partial x_1} = -c_1$, $\frac{\partial \alpha_2}{\partial x_1} = -1 - c_1 c_2$, and $\frac{\partial \alpha_2}{\partial x_2} = -c_1 - c_2$.

The final control v_0 is chosen as

$$v_0 = \alpha_n - \psi(\bar{x}) - \hat{\theta}_0^T \phi(\bar{x}) \tag{10.12}$$

$$\dot{\hat{\theta}}_0 = \Gamma\phi(\bar{x}^q)z_n \tag{10.13}$$

where Γ is a positive definite matrix.

Define an estimation error as $\tilde{\theta}_0 = \theta - \hat{\theta}_0$. Considering the Lyapunov function

$$V_0 = \sum_{i=1}^{n}\frac{1}{2}z_i^2 + \frac{1}{2}\tilde{\theta}_0^T\Gamma^{-1}\tilde{\theta}_0 \tag{10.14}$$

whose derivative can be computed as

$$\dot{V}_0 = -\sum_{i=1}^{n-1}c_iz_i^2 + z_n\left(\alpha_n + \tilde{\theta}_0^T\phi(\bar{x}) + z_{n-1}\right) - \tilde{\theta}_0^T\Gamma^{-1}\dot{\hat{\theta}}_0$$

$$\leq -\sum_{i=1}^{n}c_iz_i^2 \tag{10.15}$$

Thus it is ensured that all the signals are uniformly bounded.

10.3.2 System with Input and State Quantization

When state $x_i(t)$ is quantized with the quantizer $Q_1(x_i)$, the quantized state is defined as

$$x_i^q = Q_1(x_i), \ i = 1, 2, \ldots, n \tag{10.16}$$

Thus $\bar{x}^q(t)$ in (10.2) satisfies that $\bar{x}^q(t) = [x_1^q, \ldots, x_n^q]^T$. To facilitate the adaptive controller design, the quantized input $u(t)$ is factored into the following form

$$u(t) = Q_2(v(t)) = v(t) + d_u(t) \tag{10.17}$$

where $d_u = u(t) - v(t) \in \Re^1$. $v(t)$ denotes the control input to be designed by using the quantized states $\bar{x}^q$. Due to the property of considered quantizer in (10.6), the nonlinear part $d_u(t)$ satisfies the property given in the lemma below.

Lemma 10.1
The input quantizer error satisfies the following sector-bounded property.

$$|d_u(t)| \leq \delta_2|v(t)| + \Delta_2 \tag{10.18}$$

where $d_u(t) = u(t) - v(t)$, $u(t) = Q_2(v(t))$, $0 < \delta_2 < 1$ and $\Delta_2 > 0$ are positive quantization parameters.

Define the error variables as

$$z_1^q = x_1^q \tag{10.19}$$

$$z_i^q = x_i^q - \alpha_{i-1}^q, \ i = 2, \ldots, n \tag{10.20}$$

where

$$\alpha_1^q = -c_1 z_1^q \tag{10.21}$$

$$\alpha_i^q = -c_i z_i^q - z_{i-1}^q + \sum_{k=1}^{i-1} \frac{\partial \alpha_{i-1}}{\partial x_k} x_{k+1}^q, i = 2, \ldots, n \tag{10.22}$$

where c_i are positive constants. The adaptive controller is designed as

$$
\begin{aligned}
u(t) &= Q_2(v) & (10.23) \\
v(t) &= \alpha_n^q - \psi(\bar{x}^q) - \hat{\theta}^T \phi(\bar{x}^q) \\
&= -c_n z_n^q - z_{n-1}^q + \sum_{k=1}^{n-1} \frac{\partial \alpha_{n-1}}{\partial x_k} x_{k+1}^q \\
&\quad -\psi(\bar{x}^q) - \hat{\theta}^T \phi(\bar{x}^q) & (10.24) \\
\dot{\hat{\theta}} &= Proj\{\Gamma \phi(\bar{x}^q) z_n^q\} & (10.25)
\end{aligned}
$$

where $\hat{\theta}$ is the parameter estimator introduced for unknown vector θ, Γ is a positive definite matrix, $Proj\{.\}$ is the projector operator given in Appendix C. Note that similar to [146], the partial derivatives $\frac{\partial \alpha_{n-i}}{\partial x_k}$, $i = 1, \ldots, n$, which are calculated from functions α_i designed as (10.11) for the case when states and input are not quantized, are adopted to design α_i^q in (10.22).

Remark 10.1 The projector operator $Proj\{.\}$ is used in the parameter estimator (10.25) to ensure that the boundedness of θ, $\hat{\theta}$ and $\tilde{\theta}$ and the following property

$$-\tilde{\theta}^T \Gamma^{-1} Proj\{\tau\} \leq -\tilde{\theta}^T \Gamma^{-1} \tau, \forall \hat{\theta} \in C_\theta, \theta \in C_\theta. \tag{10.26}$$

Let $\tilde{\theta} = \theta - \hat{\theta}$, it is also ensured that $\| \tilde{\theta} \| \leq L_\theta$ and $\| \hat{\theta} \| \leq L_\theta$.

10.4 Stability Analysis

To analyze the closed-loop system stability, we first present some preliminary results in the form of the following lemmas. The proofs are given in Lemma 9.2 and Lemma 9.1 in Chapter 9.

Lemma 10.2
The state $\bar{x} = [x_1, \ldots, x_n]^T$, $\bar{x}^q = [x_1^q, \ldots, x_n^q]^T$, and the control $v(t)$ in (10.24) satisfy the following inequalities:

$$
\begin{aligned}
\| \bar{x} \| &\leq L_x \| z \| & (10.27) \\
\| \bar{x}^q \| &\leq (1 + \delta_1) L_x \| z \| + \Delta_1 & (10.28) \\
|v| &\leq L_v \| z^q \| & (10.29)
\end{aligned}
$$

where $z = [z_1, \ldots, z_n]^T$, $z^q = [z_1^q, \ldots, z_n^q]$. L_x and L_v are positive constants defined as follows, which relate to the design parameters $c_1, \ldots, c_n$.

$$L_x = \left(\sum_{i=1}^{n} L_{x_i}^2 \right)^{\frac{1}{2}} \tag{10.30}$$

$$L_v = L_{\alpha_n} + L_\psi L_x + L_\theta L_\phi L_x, \tag{10.31}$$

where

$$L_{\alpha_1} = c_1 \tag{10.32}$$

$$L_{x_i} = 1 + L_{\alpha_{i-1}}, \quad i = 2, \ldots, n \tag{10.33}$$

$$L_{\alpha_i} = c_i + 1 + \sum_{k=1}^{i-1} \left| \frac{\partial \alpha_{i-1}}{\partial x_k} \right| L_{x_i}, i = 2, \ldots, n. \tag{10.34}$$

Lemma 10.3

The effects of state quantization are bounded by functions of z as follows:

$$|\psi(\bar{x}^q) - \psi(\bar{x})| \leq \Delta_1 \Delta_\psi^1 + \delta_1 \Delta_\psi^2 \| z \| \tag{10.35}$$

$$\| \phi(\bar{x}^q) - \phi(\bar{x}) \| \leq \Delta_1 \Delta_\phi^1 + \delta_1 \Delta_\phi^2 \| z \| \tag{10.36}$$

$$|z_i^q - z_i| \leq \Delta_1 \Delta_{z_i}^1 + \delta_1 \Delta_{z_i}^2 \| \bar{z}_i \|, \quad i = 1, \ldots, n \tag{10.37}$$

$$|\alpha_i^q - \alpha_i| \leq \Delta_1 \Delta_{\alpha_i}^1 + \delta_1 \Delta_{\alpha_i}^2 \| \bar{z}_i \|, \quad i = 1, \ldots, n \tag{10.38}$$

$$\| z^q - z \| \leq \Delta_1 \Delta_z^1 + \delta_1 \Delta_z^2 \| z \| \tag{10.39}$$

where $\bar{z}_i = [z_1, \ldots, z_i]^T$, and

$$\Delta_\psi^1 = \Delta_\psi^2 = L_\psi \tag{10.40}$$

$$\Delta_\phi^1 = \Delta_\phi^2 = L_\phi \tag{10.41}$$

$$\Delta_{z_1}^1 = \Delta_{z_1}^2 = 1 \tag{10.42}$$

$$\Delta_{\alpha_1}^1 = \Delta_{\alpha_1}^2 = c_1 \tag{10.43}$$

$$\Delta_{z_i}^1 = 1 + \Delta_{\alpha_{i-1}}^1, \quad i = 2, \ldots, n \tag{10.44}$$

$$\Delta_{z_i}^2 = L_{x_i} + \Delta_{\alpha_{i-1}}^2, \quad i = 2, \ldots, n \tag{10.45}$$

$$\Delta_{\alpha_i}^1 = c_i \Delta_{z_i}^1 + \Delta_{z_{i-1}}^1 + \sum_{k=1}^{i-1} \left| \frac{\partial \alpha_{i-1}}{\partial x_k} \right|, i = 2, \ldots, n \tag{10.46}$$

$$\Delta_{\alpha_i}^2 = c_i \Delta_{z_i}^2 + \Delta_{z_{i-1}}^2 + \sum_{k=1}^{i-1} \left| \frac{\partial \alpha_{i-1}}{\partial x_k} \right| L_{x_k}, i = 2, \ldots, n \tag{10.47}$$

$$\Delta_z^j = \left(\sum_{i=1}^{n} (\Delta_{z_i}^j)^2 \right)^{1/2}, j = 1, 2 \tag{10.48}$$

Based on Lemmas 10.1, 10.2, and 10.3, the main results of this chapter can be formally stated in the following theorem.

Theorem 10.1

Consider the closed-loop system consisting of system (10.1) with state quantization (10.2) and input quantization (10.3) satisfying the property (10.6), the designed adaptive controller (10.23)–(10.24) and parameter estimator update law (10.25). If the design and quantization parameters satisfy

$$c - \delta_2\beta_1 - \delta_1\beta_2 - \sum_{i=1}^{5} r_i \geq \epsilon > 0, \tag{10.49}$$

where

$$
\begin{aligned}
c &= \min\{c_1, c_2, \ldots c_{n-1}, c_n\} & (10.50)\\
\beta_1 &= L_v(1 + \delta_1\Delta_z^2) & (10.51)\\
\beta_2 &= \Delta_{\alpha_n}^2 + \Delta_\psi^2 + B_2, & (10.52)
\end{aligned}
$$

$\epsilon, r_i(i = 1, \ldots, 5)$ *are positive constants. The following results can be guaranteed.*

1. *All the closed-loop signals are uniformly bounded.*

2. *The stabilization error $\| z(t) \|$ is ultimately bounded as follows*

$$\| z(t) \| \leq \sqrt{\frac{M}{\epsilon}} \tag{10.53}$$

where

$$
M = \frac{(\Delta_2)^2}{4r_1} + \frac{(\delta_2 L_v \Delta_1 \Delta_z^1)^2}{4r_2} + \frac{(\Delta_1 \Delta_{\alpha_n}^1)^2}{4r_3}
$$
$$
+ \frac{(\Delta_1 \Delta_\psi^1)^2}{4r_4} + \frac{(B_1)^2}{4r_5} + B_0 \tag{10.54}
$$

with

$$
\begin{aligned}
B_0 &= L_\theta(\Delta_1)^2 \Delta_\phi^1 \Delta_{z_n}^1 & (10.55)\\
B_1 &= L_\theta\Delta_1\Delta_\phi^1 + L_\theta L_\phi\Delta_1\delta_1\Delta_{z_n}^2 + L_\theta L_\phi(1 + \delta_1)L_x\Delta_1\Delta_{z_n}^1 & (10.56)\\
B_2 &= L_\theta\Delta_\phi^2 + L_\theta L_\phi(1 + \delta_1)L_x\Delta_{z_n}^2. & (10.57)
\end{aligned}
$$

Proof 10.1   The Lyapunov function for the entire closed-loop system is defined as

$$V = \sum_{i=1}^{n} \frac{1}{2}z_i^2 + \frac{1}{2}\tilde{\theta}^T\Gamma^{-1}\tilde{\theta}, \tag{10.58}$$

where z_i is given in (10.8)–(10.9) and $\tilde{\theta} = \theta - \hat{\theta}$. From (10.23)–(10.25), the derivative of V is calculated as

$$
\begin{aligned}
\dot{V} &= -\sum_{i=1}^{n-1} c_i z_i^2 + z_{n-1} z_n - \tilde{\theta}^T \Gamma^{-1} \dot{\hat{\theta}} \\
&\quad + z_n \left(v(t) + d_u - \alpha_n + \alpha_n + \psi + \phi^T \theta - \dot{\alpha}_{n-1} \right) \\
&= -\sum_{i=1}^{n-1} c_i z_i^2 - \tilde{\theta}^T \Gamma^{-1} \dot{\hat{\theta}} + z_n \left(\alpha_n^q - \psi(\bar{x}^q) - \hat{\theta}^T \phi(\bar{x}^q) \right. \\
&\quad \left. - \alpha_n + \alpha_n + \psi + \theta^T \phi - \dot{\alpha}_{n-1} + z_{n-1} + d_u \right) \\
&= -\sum_{i=1}^{n-1} c_i z_i^2 + z_n \left(\alpha_n - \dot{\alpha}_{n-1} + z_{n-1} \right) \\
&\quad + z_n d_u + z_n \left(\alpha_n^q - \alpha_n \right) + z_n \left(\psi(\bar{x}) - \psi(\bar{x}^q) \right) \\
&\quad + z_n \left(\theta^T \phi(\bar{x}) - \hat{\theta} \phi(\bar{x}^q) \right) - \tilde{\theta}^T \Gamma^{-1} \dot{\hat{\theta}} \\
&\leq -\sum_{i=1}^{n} c_i z_i^2 + |z_n| |\alpha_n^q - \alpha_n| + |z_n| |\psi(\bar{x}) - \psi(\bar{x}^q)| \\
&\quad + |z_n d_u| + \left(\theta^T \phi(\bar{x}) z_n - \hat{\theta}^T \phi(\bar{x}^q) z_n - \hat{\theta}^T \phi(\bar{x}^q) z_n^q \right)
\end{aligned}
$$

$$(10.59)$$

Using the properties (10.5), (10.6), and (10.28) in Lemma 10.2, and (10.36) and (10.37) in Lemma 10.3, the last three terms in (10.59) satisfies the following inequality

$$
\begin{aligned}
& \theta^T \phi(\bar{x}) z_n - \hat{\theta}^T \phi(\bar{x}^q) z_n - \hat{\theta}^T \phi(\bar{x}^q) z_n^q \\
&= \theta^T \phi(\bar{x}) z_n - \theta^T \phi(\bar{x}^q) z_n + \tilde{\theta} \phi(\bar{x}^q) z_n - \tilde{\theta} \phi(\bar{x}^q) z_n^q \\
&\leq \| \theta \| |z_n| |\phi(\bar{x}) - \phi(\bar{x}^q)| + \| \tilde{\theta} \| \| \phi(\bar{x}^q) \| |z_n - z_n^q| \\
&\leq L_\theta \|z\| (\Delta_1 \Delta_\phi^1 + \delta_1 \Delta_\phi^2 \|z\|) \\
&\quad + L_\theta L_\phi \|\bar{x}^q\| (\Delta_1 \Delta_{z_n}^1 + \delta_1 \Delta_{z_n}^2 \|z\|) \\
&\leq B_0 + B_1 \| z \| + \delta_1 B_2 \| z \|^2
\end{aligned}
$$

$$(10.60)$$

where B_j, $j = 0, 1, 2$, are defined in (10.55)–(10.57).
Using (10.18) and the properties (10.29), (10.39), the term $|z_n d_u|$ in (10.59) satisfies

$$
\begin{aligned}
|z_n d_u| &\leq \Delta_2 |z_n| + \delta_2 |z_n| |v| \\
&\leq \Delta_2 |z_n| + \delta_2 |z_n| L_v \| z^q \| \\
&\leq \Delta_2 |z_n| + \delta_2 |z_n| L_v (\| z \| + \Delta_1 \Delta_z^1 + \delta_1 \Delta_z^2 \| z \|) \\
&\leq \delta_2 L_v (1 + \delta_1 \Delta_z^2) \| z \|^2 + (\Delta_2 + \delta_2 L_v \Delta_1 \Delta_z^1) \|z\| \quad (10.61)
\end{aligned}
$$

Using the properties (10.35), (10.38), (10.60), (10.61) and the Young's inequality with positive parameter r (i.e. $|ab| \leq ra^2 + \frac{b^2}{4r}$), (10.59) is further computed as

$$
\begin{aligned}
\dot{V} \leq\ & -\sum_{i=1}^{n} c_i z_i^2 + \delta_2 L_v (1 + \delta_1 \Delta_z^2) \parallel z \parallel^2 + (\Delta_2 + \delta_2 L_v \Delta_1 \Delta_z^1) \parallel z \parallel \\
& + \Delta_1 \Delta_{\alpha_n}^1 |z_n| + \delta_1 \Delta_{\alpha_n}^2 \parallel z \parallel^2 + \Delta_1 \Delta_\psi^1 |z_n| + \delta_1 \Delta_\psi^2 \parallel z \parallel^2 \\
& + B_0 + B_1 \parallel z \parallel + \delta_1 B_2 \parallel z \parallel^2 \\
\leq\ & -\left(c - \delta_2 \beta_1 - \delta_1 (\Delta_{\alpha_n}^2 + \Delta_\psi^2 + B_2) - \sum_{i=1}^{5} r_i \right) \times \parallel z(t) \parallel^2 + M \\
=\ & -\left(c - \delta_2 \beta_1 - \delta_1 \beta_2 - \sum_{i=1}^{5} r_i \right) \parallel z(t) \parallel^2 + M \\
\leq\ & -\epsilon \parallel z(t) \parallel^2 + M
\end{aligned}
\tag{10.62}
$$

where the inequality $c - \delta_2 \beta_1 - \delta_1 \beta_2 - \sum_{i=1}^{5} r_i \geq \epsilon > 0$ in (10.49) has been used. ϵ, $r_i (i = 1, \ldots, 5)$ are positive constants and c, β_1, β_2, and M are defined in (10.50), (10.51), (10.52), (10.54), respectively. It is shown from (10.62) that $\dot{V} < 0$, $\forall \parallel z(t) \parallel > \sqrt{\frac{M}{\epsilon}}$. Thus the ultimate bound of $z(t)$ satisfies (10.53).

From (10.39) and the boundedness of z, z^q is bounded. Thus x_1^q and α_1^q in (10.21) is bounded. From (10.20), x_2^q is bounded. Thus α_2^q is bounded. By the same token, the boundedness of x^q and α_i^q for $i = 1, \ldots, n$ can be shown. The boundedness of $\hat{\theta}$ is ensured by the projection operator (10.25) as discussed in Remark 2. Based on Assumption 1, it implies that that $v(t)$ in (10.24) is bounded. Therefore, the boundedness of all the closed-loop signals can be ensured.

Remark 10.2   Though it is naturally motivated by the fact that measurement and control signals are transmitted via a common network in networked systems, only a few results have been developed for linear systems with both quantizations [7, 8, 17, 82]. So far, no result is available for uncertain nonlinear systems. The compensation is a non-trivial work compared to the exsiting results investigating input quantization solely. The main challenge and key technique to design adaptive backstepping controller for uncertain higher-order systems with quantized states are discussed in Chapter 9. Note that the similar technique to handle state quantization is presented in [146], where input quantization is not considered and the state quantizer satisfies error bounded property. That is, the quantization error is upper bounded by a constant automatically. Different from this, the quantization errors for sector bounded quantizers considered in this chapter depend on the quantizer inputs as appeared in (10.6).

Due to the property (10.18), the effect of quantized input cannot be simply treated as a bounded disturbance term. The key step is to establish the relation between the input signal v and the system state z by showing the property (10.29) in Lemma 10.2 and (10.39) in Lemma 10.3. To handle the issue that only quantized states $[x_1^q, \ldots, x_n^q]^T$ are used in control design, another challenge in stability analysis is to compensate for the effects of the terms $z_n(\alpha_n^q - \alpha_n)$, $z_n(\psi(\bar{x}) - \psi(\bar{x}^q))$ and $(\theta^T \phi(\bar{x})z_n - \hat{\theta}^T \phi(\bar{x}^q)z_n - \tilde{\theta}^T \phi(\bar{x}^q)z_n^q)$ in (10.59). By establishing the properties (10.27), (10.28) in Lemma 10.2 and (10.35)–(10.38) in Lemma 10.3, these terms can be bounded by the functions related to the state z. Thus all these effects of the last six terms in (10.59) can be compensated as shown in (10.62) if the condition (10.49) is satisfied.

Remark 10.3   Note that for the given design parameters c_i, the choice of quantization parameters δ_1 and δ_2 is arbitrary, provided that the inequality (10.49) holds. Therefore, (10.49) provides some insights on how to choose these quantization parameters. Besides, Theorem 10.1 implies that the number of quantization levels for δ_i is finite since all the closed-loop signals are bounded. It is noted that β_1 and β_2 are computable from the definitions (10.51) and (10.52), which depends on the design parameters c_i, the system parameters in Lemma 10.2 and Lemma 10.3, and L_ψ, L_ϕ which are assumed to be known in Assumption 10.1.

Remark 10.4   From (10.54) and (10.55)–(10.56), the upper bound of the stabilization error in the sense of (10.53) can be decreased if the quantization parameters δ_i and Δ_i are decreased while all design parameters c_i are kept unchanged.

10.5   Discussion on the Case of Bounded Quantizers

If the quantization parameter δ_i in (10.6) is set as $\delta_i = 0$, the quantizers Q_1 and Q_2 are changed to bounded quantizers having the following property:

$$|Q_i(y) - y| \ \leq \ \Delta_i, \ \ i = 1, 2 \tag{10.63}$$

In Chapter 3, it is shown that uniform quantizer and hysteresis-uniform quantizer and logarithmic-hysteresis quantizer satisfy the property in (10.63). By adopting the adaptive backstepping control algorithm presented in Section 10.3, the following results can be obtained.

Theorem 10.2
Consider the closed-loop system composed of system (10.1) with state quantization (10.2) and input quantization (10.3) satisfying the property (10.63), the designed adaptive backstepping controller (10.23)–(10.24) and parameter update law (10.25), the following results can be established.

- ■ *All the closed-loop signals are uniformly bounded.*

- ■ *The stabilization error $\| z(t) \|$ is ultimately bounded as follows*

$$\| z(t) \| \leq \sqrt{\frac{2M}{c}} \tag{10.64}$$

where c is defined in (10.50) and

$$
\begin{aligned}
M &= \frac{5}{2c}\Big[(\Delta_2)^2 + (\Delta_{\alpha_n}^1)^2 + (\Delta_\psi^1)^2 + (L_\theta \Delta_\phi^1)^2 \\
&\quad + \big(L_\theta L_\phi L_x \Delta_{z_n}^1 \big)^2 \Big] + L_\theta L_\phi \Delta_1 \Delta_{z_n}^1.
\end{aligned}
\tag{10.65}
$$

Proof 10.2 By following the analysis in the proofs of Lemma 10.2 and Lemma 10.3 with $\delta_i = 0$, the following inequalities can be obtained.

$$
\begin{aligned}
\| \bar{x} \| &\leq L_x \| z \| \tag{10.66} \\
\| \bar{x}^q \| &\leq L_x \| z \| + \Delta_1 \tag{10.67} \\
|\psi(\bar{x}^q) - \psi(\bar{x})| &\leq \Delta_\psi^1 \tag{10.68} \\
\| \phi(\bar{x}^q) - \phi(\bar{x}) \| &\leq \Delta_\phi^1 \tag{10.69} \\
|z_i^q - z_i| &\leq \Delta_{z_i}^1, \quad i = 1, \dots, n \tag{10.70} \\
|\alpha_i^q - \alpha_i| &\leq \Delta_{\alpha_i}^1, \quad i = 1, \dots, n \tag{10.71}
\end{aligned}
$$

where L_x is defined in (10.30), $\Delta_\psi^1, \Delta_\phi^1, \Delta_{z_i}^1, \Delta_{\alpha_i}^1$ are defined the same as in Lemma 10.3.

We decompose the quantized input $u(t)$ into the following two parts.

$$
u(t) = Q_2(v(t)) = v(t) + d_u(t) \tag{10.72}
$$

where $d_u = u(t) - v(t) \in \Re^1$. From (10.63), $d_u(t)$ is bounded as follows.

$$
|d_u(t)| \leq \Delta_2 \tag{10.73}
$$

Considering the Lyapunov function as defined in (10.58), its derivative can be obtained by following the control design in (10.23)–(10.25),

$$
\begin{aligned}
\dot{V} &= -\sum_{i=1}^{n} c_i z_i^2 + z_n d_u - \tilde{\theta}^T \Gamma^{-1} \dot{\hat{\theta}} + z_n \Big(\alpha_n^q - \alpha_n \Big) \\
&\quad + z_n \Big(\psi(\bar{x}) - \psi(\bar{x}^q) \Big) + z_n \Big(\theta^T \phi(\bar{x}) - \hat{\theta} \phi(\bar{x}^q) \Big) \\
&\leq -\sum_{i=1}^{n} c_i z_i^2 + |z_n d_u| + \Delta_{\alpha_n}^1 |z_n| + \Delta_\psi^1 |z_n| \\
&\quad + \Big(\theta^T \phi(\bar{x}) z_n - \hat{\theta}^T \phi(\bar{x}^q) z_n - \tilde{\theta}^T \phi(\bar{x}^q) z_n^q \Big)
\end{aligned}
\tag{10.74}
$$

where the properties (10.68) and (10.71) are used.

From (10.5), (10.63), (10.69), (10.70), and (10.67), the last three terms in (10.74) is computed as

$$
\begin{aligned}
&\theta^T \phi(\bar{x}) z_n - \hat{\theta}^T \phi(\bar{x}^q) z_n - \tilde{\theta}^T \phi(\bar{x}^q) z_n^q \\
=\ & \left(\theta^T \phi(\bar{x}) z_n - \theta \phi(\bar{x}^q) z_n \right) + \left(\tilde{\theta} \phi(\bar{x}^q) z_n - \tilde{\theta} \phi(\bar{x}^q) z_n^q \right) \\
\leq\ & \| \theta \| \, |z_n| \Delta_\phi + \| \tilde{\theta} \| \| \phi(\bar{x}^q) \| \, \Delta_{z_n}^1 \\
\leq\ & L_\theta \Delta_\phi^1 |z_n| + L_\theta L_\phi \| \bar{x}^q \| \, \Delta_{z_n}^1 \\
\leq\ & L_\theta \Delta_\phi^1 |z_n| + L_\theta L_\phi L_x \Delta_{z_n}^1 \, \| z \| + L_\theta L_\phi \Delta_1 \Delta_{z_n}^1 \quad (10.75)
\end{aligned}
$$

From (10.73), we have

$$
|z_n d_u| \leq \Delta_2 |z_n|. \tag{10.76}
$$

Then using (10.75) and (10.76) and the Young's inequality, (10.74) can be further derived as

$$
\begin{aligned}
\dot{V} \ \leq\ & -\sum_{i=1}^{n} c_i z_i^2 + \Delta_2 |z_n| + \Delta_{\alpha_n}^1 |z_n| + \Delta_\psi^1 |z_n| \\
& + L_\theta \Delta_\phi^1 |z_n| + L_\theta L_\phi L_x \Delta_{z_n}^1 \, \| z \| + L_\theta L_\phi \Delta_1 \Delta_{z_n}^1 \\
\leq\ & -\sum_{i=1}^{n} c_i z_i^2 + \frac{c}{2} \| z(t) \|^2 + M \\
\leq\ & -\frac{c}{2} \| z(t) \|^2 + M \quad (10.77)
\end{aligned}
$$

where c and M are positive constants given in (10.50) and (10.65). The inequality (10.77) implies that $\dot{V} < 0,\ \forall\ \| z(t) \| > \sqrt{\frac{2M}{c}}$. Thus the ultimate bound of $z(t)$ satisfies (10.64).

Remark 10.5   From (10.75), it can be seen that the projection technique adopted in the adaptive law (10.25) in essential to ensure the the boundedness of $\hat{\theta}$. Hence, the coupling term $\| \tilde{\theta} \| \| \phi(\bar{x}^q) \| \Delta_{z_n}^1$ can be bounded by a linear function of $\| z \|$ as shown in the last inequality of (10.75). Then Theorem 10.2 is shown without the need of a sufficient condition related to the adaptive gain and quantization parameter.

10.6   Simulation Results

We consider a pendulum system with both state and input quantization as follows.

$$
\begin{aligned}
ml\ddot{\vartheta} + mg\sin(\vartheta) + kl\dot{\vartheta} &= bu(t) &(10.78) \\
u(t) &= Q_2(v) &(10.79) \\
\bar{\vartheta}^q &= Q_1(\vartheta; \dot{\vartheta}) &(10.80)
\end{aligned}
$$

where ϑ is the angle, m and l are the mass $[kg]$ and length of the robe $[m]$, g is the gravity acceleration, k denotes an unknown friction coefficient, $u(t)$ denotes an input torque, b denotes the control coefficient. $Q_1(\vartheta; \dot{\vartheta})$ and $Q_2(v)$ are state and input quantizers, respectively. System parameters are chosen as $m = 3kg$, $l = 5m$, and $g = 9.8m/s^2$. The friction coefficient is set as $k = 0.1$. The control coefficient is set as $b = 15$. Define $\theta = \frac{k}{m}$ which is unknown system parameter.

The objective is to design v, as the input of quantizer Q_1, by using only the quantized states $\bar{\vartheta}^q$ such that the closed-loop system is stable. The adaptive controller (10.23)–(10.24) and parameter update law (10.25) are adopted. In the simulation, the initials are set as $\vartheta(0) = \dot{\vartheta}(0) = 0.8$ and $\dot{\theta}(0) = 0$. Both the input signal v and states ϑ, $\dot{\vartheta}$ are quantized by the logarithmic quantizer in (3.2) with quanzation parameters $\delta = 0.07$ and $y_{\min} = 0.1$. To validate the results stated in Theorem 10.1 and Theorem 10.2, the following two cases are considered.

- **Case 1**: Both the input signal v and states ϑ, $\dot{\vartheta}$ are quantized by a logarithmic quantizer q_l with quanzation parameters $\delta = 0.2$ and $\Delta = 0.02$. The design parameters are chosen as $c_1 = c_2 = 1.25$ and $\Gamma = 1$. With the design and quantization parameters, the condition (10.49) is satisfied. The performance of ϑ, $\dot{\vartheta}$ and the quantized states $q_l(\vartheta)$, $q_l(\dot{\vartheta})$ with respect to the time is shown in Figure 10.4. Figure 10.5 shows the designed input $v(t)$ and the quantized input $q_l(v)$.

- **Case 2**: Both the input signal v and states ϑ, $\dot{\vartheta}$ are quantized by a uniform quantizer q_u with quanzation parameter $l = 0.1$. The design parameters are chosen as $c_1 = c_2 = 4$ and $\Gamma = 1$. The performance of ϑ, $\dot{\vartheta}$ and the quantized states $q_u(\vartheta)$, $q_u(\dot{\vartheta})$ with respect to the time is shown in Figure 10.2. Besides, Figure 10.3 presents the designed input $v(t)$ and the quantized input $q_u(v)$.

Clearly, all the states and input signals are bounded. Thus the effectiveness of our proposed control scheme has been verified by the simulation results.

10.7 Notes

In this chapter, adaptive backstepping controllers are designed for uncertain nonlinear systems with both state and input quantization, where sector-bounded quantizers are considered. The projection technique is adopted to modify the design of the adaptation law. It is ensured that all the closed-loop signals are uniformly bounded and the stabilization error is ultimately bounded.

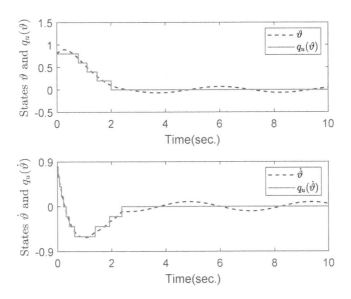

Figure 10.2 The states ϑ, $\dot{\vartheta}$ and the quantized states $Q_1(\vartheta)$, $Q_1(\dot{\vartheta})$ in Case 1

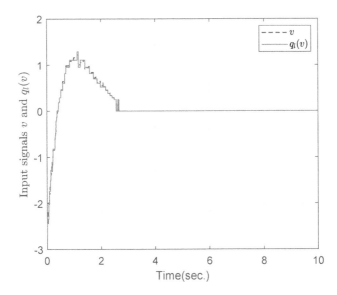

Figure 10.3 The designed input $v(t)$ and the quantized input $Q_2(v)$ in Case 1

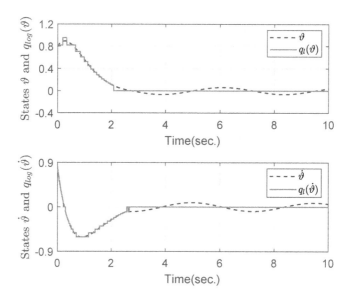

Figure 10.4 The states ϑ, $\dot{\vartheta}$ and the quantized states $Q_1(\vartheta)$, $Q_1(\dot{\vartheta})$ in Case 2

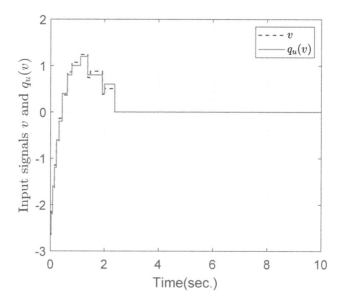

Figure 10.5 The designed input $v(t)$ and the quantized input $Q_2(v)$ in Case 2

Chapter 11

Adaptive Output Feedback Control of Systems with Input and Output Quantization

In Chapter 10, state-feedback control scheme is proposed for uncertain nonlinear systems with state and input quantization, where all the states are measurable and thus available for controller design. However, for many practical systems, only output signals can be measured. For this class of systems, output-feedback control is needed. In this chapter, an adaptive output-feedback control problem is studied for a class of uncertain nonlinear systems with both output and input quantized by a logarithmic-uniform quantizer for the sake of reducing communication burden. A control law with an adaptive gain is developed to compensate for the quantization errors. It is proved that the proposed scheme ensures that all the closed-loop signals are globally bounded. In addition, the output signal can be regulated to a bounded compact set which is explicitly given.

11.1 Introduction

It is noted that the results introduced in the previous chapters consider either only input quantization or input and state quantization. Therefore, they cannot be applied to the cases where the state signal are unavailable and output signals are quantized.

DOI: 10.1201/9781003176626-11

165

The main difficulty encountered lies in that their proposed schemes need differentiating the state and output signals to obtain the controllers, but these signals become discontinuous and thus non-differentiable after quantization. Moreover, proper state observers need to be designed to facilitate the controller design and quantization error compensation. In this chapter, we consider the output feedback control problem for a class of strict feedback non-linear systems similar to [57]. Both the output signal and the input signal are quantized by the logarithmic-uniform quantizer presented in Chapter 3. The quantized input and output signals and are utilized to design the state observer, which avoids the differentiability of discontinuous signals. But such design results in quantization errors. Subsequently, a new controller with an adaptive gain is designed to compensate for the effects of the quantization errors. It is proved that the presented adaptive controller ensures that all the closed-loop signals are globally bounded, and the output signal can be regulated to a compact set around the origin.

11.2 Problem Formulation

We consider a class of uncertain nonlinear systems in the following form.

$$\dot{x}_1 = x_2 + \phi_1(t, x)$$
$$\dot{x}_2 = x_3 + \phi_2(t, x)$$
$$\vdots \quad \vdots$$
$$\dot{x}_n = q_{lu}(u) + \phi_n(t, x)$$
$$y = x_1(t) \qquad\qquad\qquad (11.1)$$

where $x_1, \ldots, x_n$ are system states, y is the output, and u is the control input signal to be designed. $\phi_i(t, x) \in \mathbb{R} \times \mathbb{R}^n$, $i = 1, \ldots, n$ are unknown smooth nonlinear functions. $q_{lu}(u)$ denotes the output of the logarithmic-uniform quantizer and takes quantized values, which is modeled as below.

$$q_{lu}(u) = \begin{cases} q_l(u_{th}) + \lfloor \frac{u - u_{th}}{\varpi} + \kappa \rfloor \varpi, & u \geq u_{th} \\ q_l(u), & 0 \leq u < u_{th} \\ -q_s(-u), & u < 0 \end{cases} \qquad (11.2)$$

where $q_l(\cdot)$ denotes the output of a logarithmic quantizer, u_{th} is a positive constant, $\varpi = |q_l(u_{th}) - u_{th}|$, $\kappa = 1$ if $q_l(u_{th}) < u_{th}$, and otherwise $\kappa = 0$. For quantizer (11.2), we have

$$|q_{lu}(u) - u| \leq \begin{cases} \varpi & |u| \geq u_{th} \\ \delta|u| + (1 - \delta)m & |u| < u_{th} \end{cases} \qquad (11.3)$$

where $0 < \delta < 1$ and $m > 0$ are the parameters of the logarithmic quantizer. Therefore, the quantization error $|q_s(u) - u|$ is always bounded for any u and thus

$$|q_{lu}(u) - u| \leq d \qquad\qquad\qquad (11.4)$$

where d is an upper bound for the quantization error. Clearly, the logarithmic-uniform quantizer combines the traditional logarithmic and uniform quantizers to provide a balance between the system performance and communication constraints.

For system (11.1), we aim to propose a suitable controller with quantized output as a feedback signal so that all the closed-loop signals are ensured globally bounded. To achieve this control objective, the following assumption is made.

Assumption 11.1 *There is an unknown constant $c \geq 0$ such that*

$$|\phi_i(t, x)| \leq c(|x_1| + \cdots + |x_i|), \quad i = 1, \ldots, n \tag{11.5}$$

Remark 11.1   Note that Assumption 11.1 is a commonly used assumption, see [57] and [89] for examples. Many practical systems satisfy the Assumption 11.1, such as some robot-arm systems in [89].

11.3   Design of Adaptive Output Feedback Controller

In this section, we first propose a new scheme for designing a state observer and controller with an adaptive gain, in order to solve the formulated problem. Since some states are not measurable, we need to design a state observer utilizing the information of the output signal y. In our case, only the quantized output signal is available. Note that, for the input and output signal, they can be quantized by logarithmic-uniform quantizers with different parameters. For simplicity of presenting our ideas, yet without loss of generality, we assume they adopt the same parameters. Here, $q_{lu}(y)$, instead of the output signal y, is used to design the observer and controller. The adaptive controller and the state observer is designed as

$$\dot{\hat{x}}_1 = \hat{x}_2 + La_1(q_{lu}(y) - \hat{x}_1)$$
$$\dot{\hat{x}}_2 = \hat{x}_3 + L^2 a_2(q_{lu}(y) - \hat{x}_1)$$

$$\vdots \quad \vdots$$

$$\dot{\hat{x}}_n = q_{lu}(u) + L^n a_n(q_{lu}(y) - \hat{x}_1) \tag{11.6}$$

$$\dot{L} = \max\{\frac{g}{L^p}(\frac{(q_{lu}(y) - \hat{x}_1)^2}{L^2} + \frac{\hat{x}_1^2}{L^2} + \frac{\bar{d}^2}{L^2}) - \sigma, 0\}, \quad L(0) = 1 \tag{11.7}$$

$$u = -[L^n k_1 \hat{x}_1 + L^{n-1} k_2 \hat{x}_2 + \cdots + Lk_n \hat{x}_n] - \bar{d} \tanh(\frac{z^T Q \omega_n}{\varepsilon L^n}) \tag{11.8}$$

where $z = [z_1, z_2, \ldots, z_n]^T$ and Q is a positive-definite symmetric matrix defined respectively in (11.14) and (11.18), $p \geq 1$, $g > 0$, $\bar{d} \geq d$, and $\sigma > 0$ are design parameters, ω_n is an n-dimensional column vector with the nth entry being 1 and the others being 0, and $a_i > 0$ and $k_i > 0$ $(i = 1, \ldots, n)$ are also design parameters chosen to make the following two matrices Hurwitz:

$$A = \begin{pmatrix} -a_1 & 1 & \cdots & 0 \\ \vdots & \vdots & \ddots & \vdots \\ -a_{n-1} & 0 & \cdots & 1 \\ -a_n & 0 & \cdots & 0 \end{pmatrix} \quad B = \begin{pmatrix} 0 & 1 & \cdots & 0 \\ \vdots & \vdots & \ddots & \vdots \\ 0 & 0 & \cdots & 1 \\ -k_1 & -k_2 & \cdots & -k_n \end{pmatrix} \quad (11.9)$$

Let $e_i = x_i - \hat{x}_i$ be the state estimation error. Then, the error dynamics is given as

$$\dot{e}_1 = e_2 + \phi_1 - La_1(q_{lu}(y) - \hat{x}_1)$$
$$\dot{e}_2 = e_3 + \phi_2 - L^2 a_2(q_{lu}(y) - \hat{x}_1)$$
$$\vdots \quad \vdots$$
$$\dot{e}_n = \phi_n - L^n a_n(q_{lu}(y) - \hat{x}_1) \quad (11.10)$$

Now we study $q_{lu}(y)$ by only considering the condition that $y > 0$, as the same analysis can be repeated for the case that $y \leq 0$. From (11.4) we know

$$y - d \leq q_{lu}(y) \leq y + d \quad (11.11)$$

which yields

$$q_{lu}(y) = y + \mu_1 d \quad (11.12)$$

where μ_1 can be any number in $[-1, 1]$. Similarly, for $q_{lu}(u)$ we have $q_{lu}(u) = u + \mu_2 d$ where μ_2 can be any number in $[-1, 1]$. Thus (11.10) can be transformed into

$$\dot{e}_1 = e_2 + \phi_1 - La_1 e_1 - \mu_1 La_1 d$$
$$\dot{e}_2 = e_3 + \phi_2 - L^2 a_2 e_1 - \mu_1 L^2 a_2 d$$
$$\vdots \quad \vdots$$
$$\dot{e}_n = \phi_n - L^n a_n e_1 - \mu_1 L^n a_n d \quad (11.13)$$

Now we consider the following coordinate transformations

$$\epsilon_i = \frac{e_i}{L^i} \quad \text{and} \quad z_i = \frac{\hat{x}_i}{L^i} \quad (11.14)$$

Then, based on (11.6), (11.8), and (11.13) we can obtain

$$\dot{\epsilon} = LA\epsilon - \frac{\dot{L}}{L}D\epsilon + \Phi - \mu_1 ad \quad (11.15)$$

$$\dot{z} = LBz + La\epsilon_1 - \frac{\dot{L}}{L}Dz + \mu_1 ad + \omega_n\frac{\mu_2 d}{L^n} - \omega_n\frac{\bar{d}}{L^n}\tanh(\frac{z^T Q\omega_n}{\varepsilon L^n}) \quad (11.16)$$

where $\epsilon = [\epsilon_1, \epsilon_2, \ldots, \epsilon_n]^T$, $z = [z_1, z_2, \ldots, z_n]^T$, $a = [a_1, a_2, \ldots, a_n]^T$, $D = diag\{1, 2, \ldots, n\}$, and $\Phi = [\frac{\phi_1}{L}, \frac{\phi_2}{L^2}, \ldots, \frac{\phi_n}{L^n}]^T$.

11.4 Stability Analysis

By analyzing the stability of the closed-loop system, the following theorem is established.

Theorem 11.1
Consider the closed-loop system consisting of system (11.1) under Assumption 11.1, the state observers (11.6), and the control law (11.8) with the adaptive control gain (11.7). All the closed-loop signals are globally bounded and the output signal can converge to the following compact set.

$$\Omega_y = \{y| \ |y| \le \sqrt{\frac{4L_{max}^{p+2}\sigma}{g}}\} \tag{11.17}$$

where L_{max} is the maximum value of $L(t)$.

Proof 11.1   As A and B are Hurwitz, from [57] there must exist positive-definite symmetric $P > 0$ and $Q > 0$ such that

$$A^T P + PA \le -I \quad and \quad DP + PD \ge 0$$
$$B^T Q + QB \le -2I \quad and \quad DQ + QD \ge 0 \tag{11.18}$$

Then we choose the Lyapunov function as

$$V = (m_1 + 1)\epsilon^T P\epsilon + z^T Qz \tag{11.19}$$

where $m_1 = ||Q||^2||a||^2$. From Lemma D.1 in Appendix D, we obtain that

$$\dot{V} = 2(m_1 + 1)\epsilon^T P(LA\epsilon - \frac{\dot{L}}{L}D\epsilon + \Phi - \mu_1 ad) + 2z^T Q(LBz$$

$$+ La\epsilon_1 - \frac{\dot{L}}{L}Dz + \mu_1 ad + \omega_n \frac{\mu_2 d}{L^n} - \omega_n \frac{d}{L^n}\tanh(\frac{z^T Q\omega_n}{\varepsilon L^n})) \tag{11.20}$$

Substituting (11.18) into (11.20) gives

$$\dot{V} \le -(m_1 + 1)L||\epsilon||^2 - 2L||z||^2 + 2(m_1 + 1)\epsilon^T P\Phi$$
$$- 2(m_1 + 1)\epsilon^T P\mu_1 ad + 2z^T QLa\epsilon_1 + 2z^T Q\mu_1 ad$$
$$+ 2z^T Q(\omega_n \frac{\mu_2 d}{L^n} - \omega_n \frac{d}{L^n}\tanh(\frac{z^T Q\omega_n}{\varepsilon L^n})) \tag{11.21}$$

Now we estimate the cross terms in (11.21). First by noting that $L(t) \ge 1$ all the time, we have

$$-2(m_1 + 1)\epsilon^T P\mu_1 ad \le d||\epsilon||^2 + d||(m_1 + 1)Pa||^2 \tag{11.22}$$
$$2z^T QLa\epsilon_1 \le L||z||^2 + Lm_1\epsilon_1^2 \tag{11.23}$$
$$2z^T Q\mu_1 ad \le d||z||^2 + dm_1 \tag{11.24}$$

From Assumption 11.1, we have

$$|\frac{\phi_i}{L^i}| \leq \frac{c}{L^i}(|\hat{x}_1| + \cdots + |\hat{x}_i| + |e_1| + \cdots + |e_i|)$$
$$\leq c\sqrt{n}(||z|| + ||\epsilon||) \tag{11.25}$$

which yields

$$|2(m_1 + 1)\epsilon^T P\Phi| \leq 2(m_1 + 1)cn||\epsilon|| \, ||P||(||z|| + ||\epsilon||)$$
$$\leq 3n||P||c(m_1 + 1)(||z||^2 + ||\epsilon||^2) \tag{11.26}$$

Now we need to estimate the last term of (11.21). Note that the hyperbolic tangent function $\tanh(\cdot)$ has the following property,

$$0 \leq |\varrho| - \varrho\tanh(\frac{\varrho}{\varepsilon}) \leq 0.2785\varepsilon \tag{11.27}$$

where $\varepsilon > 0$ and $\varrho \in \mathbb{R}$. As a result, we have

$$2z^T Q(\omega_n\frac{\mu_2 d}{L^n} - \omega_n\frac{\bar{d}}{L^n}\tanh(\frac{z^T Q\omega_n}{\varepsilon L^n})) \leq 0.557\varepsilon \tag{11.28}$$

Substituting (11.22)–(11.24), (11.26) and (11.28) into (11.21) yields

$$\dot{V} \leq -[L - cm(m_1 + 1) - d](||z||^2 + ||\epsilon||^2) + \Delta \tag{11.29}$$

in which $m = 3n||p||$ and $\Delta = d||(m_1 + 1)Pa||^2 + dm_1 + 0.557\varepsilon$.

The remaining proof will be divided into four parts. In the first part, we will show the adaptive gain L is bounded, while the boundedness of z and ϵ are shown in Part 2 and Part 3, respectively. The property of the output signal is analyzed in Part 4.

Without loss of generality, we first assume that $(L(t), \epsilon(t), z(t))$ has a continuous solution on the maximally extended interval $[0, t_f)$ with t_f satisfying $0 < t_f < +\infty$. At last, we will show that $t_f = +\infty$.

1. In this part, we will show that $L(t)$ will not escape at time t_f by contradiction arguments. First, we assume $\lim_{t \to t_f} L(t) = +\infty$. From (11.7), we know L is a non-decreasing function. Therefore, there must exist a finite time $t^* \in (0, t_f)$ such that

$$L(t) \geq cm(m_1 + 1) + d + 1, \quad t^* \leq t \leq t_f \tag{11.30}$$

Let

$$\alpha = \max\{(m_1 + 1)\lambda_{max}(P), \lambda_{max}(Q)\} \tag{11.31}$$
$$\beta = \min\{(m_1 + 1)\lambda_{min}(P), \lambda_{min}(Q)\} \tag{11.32}$$

where $\lambda_{max}(P)$ and $\lambda_{min}(P)$ denote the biggest and smallest eigenvalues of matrix P, respectively. The same notation applies for matrix Q. From (11.29) and (11.30), it follows that

$$\dot{V} \leq -\frac{1}{\alpha}V + \Delta \quad t^* \leq t \leq t_f \tag{11.33}$$

Consequently, for $t^* \leq t \leq t_f$, V must be bounded. Moreover, we can get

$$V(t) \leq e^{-\frac{1}{\alpha}(t-t^*)} V(t^*) + \alpha\Delta(1 - e^{-\frac{1}{\alpha}(t-t^*)}) \leq V(t^*) + \alpha\Delta \qquad (11.34)$$

Then, from (11.7) we have for $t^* \leq t \leq t_f$

$$
\begin{aligned}
& \frac{g}{L(t)^p}\left(\frac{(q_{lu}(y) - \hat{x}_1)^2}{L(t)^2} + \frac{\hat{x}_1^2}{L(t)^2} + \frac{\bar{d}^2}{L(t)^2}\right) \\
\leq \ & \frac{g}{L(t)^p}\left(\frac{2e_1^2 + 2u_1^2 d^2}{L(t)^2} + \frac{\hat{x}_1^2}{L(t)^2} + \frac{\bar{d}^2}{L(t)^2}\right) \\
\leq \ & \frac{2g}{L(t)^p \beta}(V(t^*) + \alpha\Delta) + \frac{g(2d^2 + \bar{d}^2)}{L(t)^{2+p}} \qquad (11.35)
\end{aligned}
$$

Since $L(t)$ is non-decreasing and $\lim_{t \to t_f} L(t) = +\infty$, there must exist a time $t_1^* \in [t^*, t_f]$ such that $\frac{2g}{L(t_1^*)^p \beta}(V(t^*) + \alpha\Delta) + \frac{g(2d^2 + \bar{d}^2)}{L(t_1^*)^{2+p}} = \sigma$. This means for all $t \in (t_1^*, t_f)$, $\dot{L}(t) = 0$, which leads to a contraction by the assumption $\lim_{t \to t_f} L(t) = +\infty$. Therefore, $L(t)$ must be bounded.

2. Now we will show that $z(t)$ is also bounded on the interval $[0, t_f)$. To this end, we choose the Lyapunov function as

$$V_z = z^T Q z \qquad (11.36)$$

Then, the derivative of V_z along (11.16) is given by

$$\dot{V}_z \leq (-L + \frac{1}{2})\|z\|^2 + m_1 L\epsilon_1^2 + 2m_1 d^2 + 0.557\varepsilon \qquad (11.37)$$

From (11.7), we know

$$
\begin{aligned}
\dot{L} &\geq \frac{g}{L^p}\left[\frac{\frac{1}{2}e_1^2 - d^2}{L^2} + \frac{\hat{x}_1^2}{2L^2} + \frac{\bar{d}^2}{L^2}\right] - \sigma \\
&= \frac{g}{2L^p}(\epsilon_1^2 + z_1^2) - \sigma \qquad (11.38)
\end{aligned}
$$

Combining (11.38) and (11.37) yields

$$
\begin{aligned}
\dot{V}_z &\leq (-L + \frac{1}{2})\|z\|^2 + \frac{2m_1 L^{p+1}}{g}\dot{L} + \frac{2m_1 L^{p+1}}{g}\sigma + 2m_1 d^2 + 0.557\varepsilon \\
&\leq -c_1 V_z + c_2 \dot{L} + c_3 \qquad (11.39)
\end{aligned}
$$

where c_1, c_2, and c_3 are constants satisfying $c_1 \geq \frac{L - \frac{1}{2}}{\lambda_{max}(Q)} > 0$, $c_2 \geq \frac{2m_1 L^{p+1}}{g}$, $c_3 \geq \frac{2m_1 L^{p+1}}{g}\sigma + 2m_1 d^2 + 0.557\varepsilon$. Integrating both sides of (11.39) gives

$$V_z \leq e^{-c_1 t} V_z(0) + c_2[L(t) - 1 - c_1 \int_0^t e^{c_1 \tau} L(\tau)d\tau] + \frac{c_3}{c_1}(1 - e^{-c_1 t}) \qquad (11.40)$$

From (11.40) we get that z will eventually converge towards the set $\Omega_z = \{z|\ ||z||^2 \leq \frac{c_2 L(t) + \frac{c_3}{c_1}}{\lambda_{min}(Q)}\}$. Since $L(t)$ is bounded, from the definition of c_i $(i = 1, 2, 3)$ and $m_1 = ||Q||^2||a||^2$, we can obtain that Ω_z must be bounded.

3. Now we prove that ϵ is also well defined in $[0, t_f)$. We consider the following change of coordinates

$$\xi_i = \frac{e_i}{(L^*)^i} \tag{11.41}$$

in which L^* is a constant satisfying

$$L^* \geq \max\{\sup\{L(t)\}, 3cn||P|| + 3\} \tag{11.42}$$

Then, from (11.13) we have

$$\dot{\xi} = L^* A\xi + L^* a\xi_1 - L\tau_1 a\xi_1 + \Phi^* - \tau_2 \mu_1 ad \tag{11.43}$$

where $\xi = [\xi_1, \ldots, \xi_n]^T$, $\tau_1 = diag\{1, \frac{L}{L^*}, \ldots, (\frac{L}{L^*})^{n-1}\}$, $\tau_2 = diag\{\frac{L}{L^*}, \ldots, (\frac{L}{L^*})^n\}$, and $\Phi^* = [\frac{\phi_1}{L^*}, \ldots, \frac{\phi_n}{(L^*)^n}]^T$. Now consider the Lyapunov function $V_\xi = \xi^T P\xi$ for system (11.43), we have

$$\begin{aligned}
\dot{V}_\xi = \dot{\tilde{V}}_\xi &= 2\xi^T P(L^* A\xi + L^* a\xi_1 - L\tau_1 a\xi_1 + \Phi^* - \tau_2 \mu_1 ad) \\
&\leq -L^*||\xi||^2 + 2\xi_1 L^* \xi^T Pa - 2\xi_1 L\xi^T P\tau_1 a \\
&\quad + 2\xi^T P\Phi^* - 2\xi^T P\tau_2 \mu_1 ad
\end{aligned} \tag{11.44}$$

Since

$$|2\xi_1 L^* \xi^T Pa| \leq (L^*)^2||a^T P||^2\xi_1^2 + ||\xi||^2 \tag{11.45}$$

$$|-2\xi_1 L\xi^T P\tau_1 a| \leq (L^*)^2||a^T P\tau_1||^2\xi_1^2 + ||\xi||^2 \tag{11.46}$$

$$|2\xi^T P\Phi^*| \leq 3n||P||c(||z||^2 + ||\xi||^2) \tag{11.47}$$

$$|2\xi^T P\tau_2 \mu_1 ad| \leq d^2||a^T P\tau_2||^2 + ||\xi||^2 \tag{11.48}$$

Substituting (11.45)–(11.48) into (11.44) yields

$$\begin{aligned}
\dot{V}_\xi &\leq -(L^* - 3nc||P|| - 3)||\xi||^2 + 3nc||P||\ ||z||^2 \\
&\quad + [(L^*)^2||a^T P||^2 + (L^*)^2||a^T P\tau_1||^2]\xi_1^2 + d^2||a^T P\tau_2||^2
\end{aligned} \tag{11.49}$$

Since $L^* \geq \sup(L(t))$, we have $|\xi_1| \leq |\epsilon_1|$. Combining (11.38) and (11.49) gives

$$\dot{V}_\xi \leq -c_4 V + c_5 \dot{L} + c_6 \tag{11.50}$$

in which c_4, c_5, and c_6 are constants satisfying

$$c_4 \geq \frac{L^* - 3nc\|P\| - 3}{\lambda_{max}(P)}$$

$$c_5 \geq \frac{2L^p}{g}[(L^*)^2\|a^T P\|^2 + (L^*)^2\|a^T P\tau_1\|^2]$$

$$c_6 \geq \frac{2L^p}{g}[(L^*)^2\|a^T P\|^2 + (L^*)^2\|a^T P\tau_1\|^2]\sigma$$

$$+ 3nc\|P\| \|z\|^2 + d^2\|a^T P\tau_2\|^2 \tag{11.51}$$

Following the same analysis for system (11.39), we obtain ξ will eventually converge to a set $\Omega_\xi = \{\xi| \|\xi\|^2 \leq \frac{c_5 L(t)+\frac{c_6}{c_4}}{\lambda_{min}(Q)}\}$ which is bounded. Since $\epsilon = (\frac{L^*}{L})^2\xi$, the boundedness of ϵ is guaranteed.

Now we have obtained that $(L(t), \epsilon(t), z(t))$ are all bounded on the maximum time interval $[0, t_f)$. Thus it can be concluded that $t_f = +\infty$, similar to [57] and [58].

4. Finally, we show to what extent we can regulate the output signal y. Since $L(t)$ is bounded and $\dot{L}(t) \geq 0$, by Barbalat's Lemma we can obtain $\lim_{t\to\infty} \dot{L}(t) = 0$. Therefore, there must exist a time t_2^* such that

$$\frac{g}{L^p}\left(\frac{(q_{lu}(y) - \hat{x}_1)^2}{L^2} + \frac{\hat{x}_1^2}{L^2} + \frac{\bar{d}^2}{L^2}\right) - \sigma \leq 0, \quad t > t_2^* \tag{11.52}$$

which leads to

$$\epsilon_1^2 + z_1^2 \leq \frac{2L^p\sigma}{g} \tag{11.53}$$

Then, we have

$$y^2 = (e_1 + \hat{x}_1)^2 \leq 2L^2\epsilon_1^2 + 2L^2 z_1^2 \leq \frac{4L^{p+2}\sigma}{g} \leq \frac{4L_{max}^{p+2}\sigma}{g} \tag{11.54}$$

Thus the output signal y will eventually converge to the bounded set $\Omega_y = \{y| |y| \leq \sqrt{\frac{4L_{max}^{p+2}\sigma}{g}}\}$.

Remark 11.2 Compared with the existing results in [57] without signal quantization, the adaptive control gain law (11.7) and the final control law (11.8) are both modified to achieve the control objective. To be more specific, the adaptive control gain law (11.7) is totally different from that in [57], which is essential to guarantee the boundedness of $L(t)$, $z(t)$, $\epsilon(t)$, as seen from (11.35), (11.38), and (11.52). For the control law, a nonlinear part, i.e. the last term in (11.8) is added to handle the input quantization errors, as shown in (11.28).

Remark 11.3   It must be pointed out that the proposed output feedback control scheme in this paper is not based on the separation principle. Specifically, we design the controller and the observer at the same time, and the obtained observer and controller are strongly coupled with each other intrinsically. As seen from our analysis, the Lyapunov function defined in (11.19) contains the system states and also estimated states. Thus their boundedness is established without considering them separately. Detailed discussions can also be found in [89].

11.5   An Illustrative Example

Now we consider an illustrating example from [57], and both the input and output signals are quantized for control.

$$\dot{x}_1 = x_2 + \frac{x_1}{(1 - c_1 x_2)^2 + x_2^2}$$
$$\dot{x}_2 = q_s(u) + \ln(1 + (x_2^2)^{c_2})$$
$$y = x_1 \qquad\qquad (11.55)$$

where c_1 and $c_2 \geq 1$ are unknown constants. Assumption 11.1 for (11.55) is satisfied since

$$\left| \frac{x_1}{(1 - c_1 x_2)^2 + x_2^2} \right| \leq (1 + c_1^2)|x_1|$$
$$\ln(1 + (x_2^2)^{c_2}) \leq (2c_2 - 1)|x_2| \qquad\qquad (11.56)$$

Note that the proposed control scheme in [57] does not take signal quantization into account. To show the effectiveness of our approach, we make comparisons of the control scheme in [57] and this paper both under the condition that the input and output signals are quantized. Specifically, we choose the same simulation parameters for the system as those in [57], i.e. $c_1 = 1$, $c_2 = 2$, $(x_1(0), x_2(0)) = (1, 5)$, $(\hat{x}_1(0), \hat{x}_2(0)) = (-10, -2)$, $a_1 = a_2 = 1$, and $k_1 = k_2 = 1$. The parameters of the quantizer are set as $m = 0.02$, $\delta = 0.2$, $u_{th} = 24.6$, and $\varpi = 4.9$. Besides, we choose the remaining parameters as $g = 1.5$, $\sigma = 0.8$, $\bar{d} = 10$, $\varepsilon = 0.01$, and $Q = \begin{pmatrix} 3 & 1 \\ 1 & 2 \end{pmatrix}$. For the closed loop system, Figures 11.1–11.2 show the state performances with the scheme in [57] and the proposed scheme in this chapter, while the adaptive control gain and the control signal are presented in Figure 11.3. From these simulation results, it is observed that the time needed for the states to converge to zero by using the control scheme in this paper and the one in [57] is around 5 seconds and 10 seconds respectively. Moreover, the amplitudes of the state oscillation and that of the control signal with our proposed scheme are also smaller. These results illustrate that the proposed design method gives better performances.

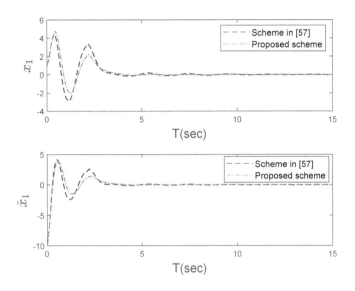

Figure 11.1 State performance x_1 and $\hat{x}_1$

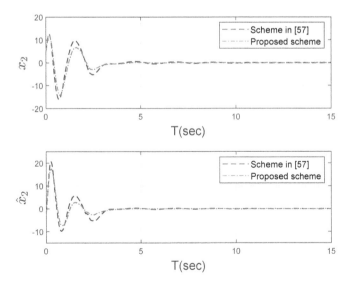

Figure 11.2 State performance x_2 and $\hat{x}_2$

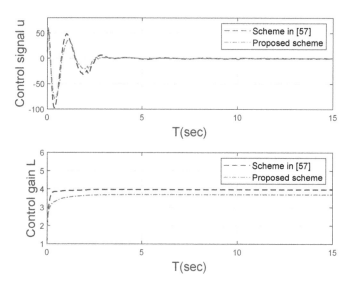

Figure 11.3 Control signal

11.6 Notes

The adaptive output feedback control is developed for a class of uncertain nonlinear systems with unknown growth rate, where the output signal and the input signal of the system are quantized at the same time. A new controller with modified adaptive control gains is proposed to compensate for the effects of the quantization errors. The proposed scheme ensures the global boundedness of all the states of the uncertain system.

Acknowledgment

Reprinted from Copyright (2021), with permission from Elsevier. Lantao Xing, Changyun Wen, Lei Wang, Zhitao Liu, Hongye Su, "Adaptive output feedback regulation for a class of nonlinear systems subject to input and output quantization", *Journal of the Franklin Institute*, vol. 354, pp. 6536–6549, 2017.

APPLICATIONS

Chapter 12

Adaptive Attitude Control of Helicopter with Quantization

This chapter proposes an adaptive controller for a helicopter system in the presence of input quantization. A nonlinear mathematical model is derived for a 2-Degree of Freedom (DOF) helicopter system based on Euler-Lagrange equations, where some system parameters are uncertain. The inputs are quantized by uniform quantizers. An adaptive control algorithm is developed by using the backstepping technique to track the pitch and yaw position references independently. Only quantized input signals are used in the system which reduces communication rate and costs. It is shown that not only the ultimate stability is guaranteed by the proposed controller, but also the designers can tune the design parameters in an explicit way to obtain the required closed-loop behavior. Experiments are carried out on the Quanser helicopter to validate the effectiveness, robustness, and control capability of the proposed scheme.

12.1 Introduction

Attitude control of rigid bodies has been widely addressed in the literature, see for examples, [2, 5, 13, 56, 90, 91, 102, 120], and with applications in marine systems in [27], unmanned aerial vehicles in [14], helicopters in [92], underwater vehicles in [128], and other robotic systems. In [56], a robust adaptive controller is proposed for the attitude tracking problem of rigid bodies in the presence of uncertain parameters and where the attitude is represented by rotation matrices. In [5], an adaptive attitude tracking controller is developed for rigid body systems in the presence of unknown inertia and gyro-bias. In [102], an adaptive controller is proposed for a leader-following attitude consensus problem for multiple rigid body systems subject to jointly connected switching networks in the presence of uncertain parameters. In [2], an adaptive backstepping controller was proposed for the trajectory tracking of a rigid body with unknown mass and inertia based on dual-quaternions. [14] proposed a robust nonlinear controller for a quadrotor unmanned aerial vehicles, which combines the sliding-mode control technique and the backstepping control technique. [128] investigated the sliding mode tracking control of underwater vehicles. In [92], adaptive backstepping control is proposed for pitch and yaw control of a 2-DOF Helicopter system.

Quantized control is a potential problem and has attracted considerable attention in recent years due to its theoretical and practical importance in practical engineering, where digital processors are widely used and signals are required to be quantized and transmitted via a common communication network. Quantized control of rigid bodies is a potential problem. Attitude stabilization with input quantization was investigated in [96] using a fixed-time sliding mode control. Trajectory tracking control for autonomous underwater vehicles with quantization was investigated in [128] using a sliding mode controller, where the considered systems are completely known. In [36], adaptive tracking control was proposed for underactuated autonomous underwater vehicles with input quantization.

Uncertainties and non-linearities always exist in many practical systems. Thus it is more reasonable to consider adaptive control of rigid body systems with quantization. There are still very few results for attitude control of rigid body system with uncertain parameters and input quantization using backstepping technique. In [93], an adaptive backstepping control scheme with quantized inputs is presented for a helicopter system, considering a uniform input quantizer. In [36], adaptive backstepping was investigated for tracking control for under-actuated autonomous underwater vehicles with input quantization. An adaptive backstepping controller was proposed for formation tracking controller for a group of UAVs with quantized inputs in [105]. This chapter is concerned with the attitude tracking control of a helicopter system with input quantization. A nonlinear mathematical model is derived for the 2-DOF helicopter system based on Euler-Lagrange equations, where some system parameters are uncertain. An adaptive control algorithm is developed by using backstepping technique to track the pitch and yaw position references independently. Only quantized input signals are used in the system which reduces communication rate and costs. It is shown that not only the ultimate stability is guaranteed by the proposed controller, but also the designers can tune the design parameters in an explicit way to obtain the required closed loop behavior. Experiments are carried out on the Quanser helicopter system to validate the effectiveness, robustness and control capability of the proposed scheme.

12.2 Problem Statement

12.2.1 System Model

The helicopter system is visualized in Figure 12.1 showing both a free body diagram (FBD) and a kinetic diagram (KD). The main motor is producing two forces, one main force, F_{Mz}, in the z_b-direction that will give a positive pitch angle, and also a force, F_{My}, in the y_b-direction, meaning this will give a yaw angle. This last force is due to the aerodynamic forces. The tail motor is also producing two forces, F_{Tz} and F_{Ty}. This motor is basically to counteract the yaw from the main motor and thus control the yaw while the main motor is controlling the pitch. These forces are functions of the two system inputs, u_1 and u_2, that are the voltages applied to the main and tail motors. Viscous damping, proportional to the velocity of the Aero body, is also present.

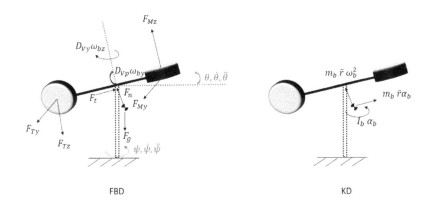

FBD   KD

Figure 12.1 Free body diagram and kinetic diagram of the Aero body

This is a MIMO system with 2 DOF, where each input will change both the pitch and yaw angle. The helicopter model is considered as a rigid body and the equations of motion are derived using Euler-Lagrange equations as given in [92], where the system parameters and control coefficients are uncertain.

The state variables are defined as

$$\boldsymbol{x} = [\vartheta(t), \psi(t), \dot{\vartheta}(t), \dot{\psi}(t)]^\top, \tag{12.1}$$

where ϑ and ψ are pitch and yaw angles, and $\dot{\vartheta}$ and $\dot{\psi}$ are angular velocities of pitch and yaw angles. The control variables are defined as

$$\boldsymbol{u} = [u_1(t, \boldsymbol{x}), u_2(t, \boldsymbol{x})]^\top, \tag{12.2}$$

and are the inputs that will be quantized. The nonlinear state space model is expressed as

$$
\dot{x} =
\begin{bmatrix}
x_3 \\
x_4 \\
\phi_1^\top \theta_1 \\
\phi_2^\top \theta_2
\end{bmatrix}
+
\begin{bmatrix}
0 \\
0 \\
\beta_{1,1} u_1 + \beta_{1,2} u_2 \\
-\beta_{2,1} u_1 + \beta_{2,2} u_2
\end{bmatrix}
\tag{12.3}
$$

where ϕ_1 and ϕ_2 are known nonlinear functions defined as

$$
\phi_1 =
\begin{bmatrix}
-x_3 \\
-\sin x_1 \\
x_4^2 \cos x_1 \sin x_1
\end{bmatrix}, \quad
\phi_2 =
\begin{bmatrix}
-x_4 \\
-x_2 x_4 \cos x_1 \sin x_1
\end{bmatrix},
\tag{12.4}
$$

vectors θ_1 and θ_2 are unknown constant vectors defined as

$$
\theta_1 = \frac{1}{I_p + ml_{cm}^2}
\begin{bmatrix}
D_{Vp} \\
mgl_{cm} \\
ml_{cm}^2
\end{bmatrix}, \quad
\theta_2 = \frac{1}{I_y}
\begin{bmatrix}
D_{Vy} \\
2ml_{cm}^2
\end{bmatrix},
\tag{12.5}
$$

and $\beta_{i,j}$, $i,j \in \{1,2\}$, are unknown constants defined as

$$
\beta_{1,1} = \frac{K_{pp}}{I_p + ml_{cm}^2}, \qquad \beta_{1,2} = \frac{K_{py}}{I_p + ml_{cm}^2}, \tag{12.6}
$$

$$
\beta_{2,1} = \frac{K_{yp}}{I_y}, \qquad \beta_{2,2} = \frac{K_{yy}}{I_y}. \tag{12.7}
$$

The constants K_{pp} and K_{yy} are torque thrust gains from main and tail motors, K_{py} is cross-torque thrust gain acting on pitch from tail motor, K_{yp} is cross-torque thrust gain acting on yaw from main motor, l_{cm} is the distance between the center of mass and the origin of the body-fixed frame, I_p and I_y are the moments of inertia of the pitch and yaw respectively, g is the gravity acceleration, m is the total mass of the Aero body, and D_{V_y} and D_{V_p} are the damping constants for the rotation along the yaw axis and pitch axis separately.

The control objective is to design a control law for u_1 and u_2 to force the outputs x_1 and x_2 to track the reference signals $x_{r1}(t)$ and $x_{r2}(t)$ for pitch and yaw respectively when the inputs are quantized. To achieve the objective, the following assumptions are imposed.

Assumption 12.1 *The reference signals x_{r1} and x_{r2} and first and second order derivatives are known, piecewise continuous and bounded.*

Assumption 12.2 *All unknown parameters θ_1, θ_2, $\beta_{i,j}$, $i,j \in \{1,2\}$ are positive constants and within known bounds.*

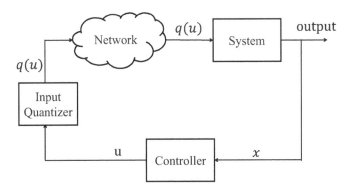

Figure 12.2 System with quantized inputs

12.2.2 *Quantized System*

In this paper, we consider a quantized feedback system as shown in Figure 12.2. The inputs u_1 and u_2 in system (12.2.1) take the quantized values, that are quantized at the encoder side. The control inputs u_1 and u_2 are quantized using a uniform quantizer which has intervals of fixed lengths and is defined as follows:

$$q(u_k) = \begin{cases} u_{k,i}\mathrm{sgn}(u_k), & u_{k,i} - \frac{l_k}{2} \leq |u_k| < u_{k,i} + \frac{l_k}{2} \\ 0, & |u_k| < u_{k,0} + \frac{l_k}{2} \end{cases} \tag{12.8}$$

where $k = 1, 2$, $u_{k,0} > -l_k/2$ is a constant, $l_k > 0$ is the length of the quantization intervals, $i = 1, 2, \ldots$, and $u_{k,i+1} = u_{k,i} + l_k$. The uniform quantization $q(u_k) \in U_k = \{0, \pm u_{k,i}\}$. The smaller the quantization intervals are, the closer the signal is to its continuous counterpart.

12.3 Adaptive Control Design

In this section, we will design adaptive feedback control laws for the helicopter model using backstepping technique. First considering the case when the inputs are continuous and then with quantized inputs.

12.3.1 Continuous Inputs

We begin by introducing the change of coordinates

$$z_1 = x_1 - x_{r1}, \tag{12.9}$$

$$z_2 = x_2 - x_{r2}, \tag{12.10}$$

$$z_3 = x_3 - \alpha_1 - \dot{x}_{r1}, \tag{12.11}$$

$$z_4 = x_4 - \alpha_2 - \dot{x}_{r2}. \tag{12.12}$$

where α_1 and α_2 are the virtual controllers. The design follows the backstepping procedure in [53].

• *Step* 1: The virtual controllers are chosen as

$$\alpha_1 = -c_1 z_1, \tag{12.13}$$

$$\alpha_2 = -c_2 z_2, \tag{12.14}$$

where c_1 and c_2 are positive constants. A control Lyapunov function is chosen as

$$V_1(\boldsymbol{z}, t) = \frac{1}{2} z_1^2 + \frac{1}{2} z_2^2. \tag{12.15}$$

The derivative of V_1 along the solutions of the system is

$$
\begin{aligned}
\dot{V}_1 &= z_1 \dot{z}_1 + z_2 \dot{z}_2 \\
&= z_1(z_3 + \alpha_1) + z_2(z_4 + \alpha_2) \\
&= -c_1 z_1^2 + z_1 z_3 - c_2 z_2^2 + z_2 z_4.
\end{aligned}
\tag{12.16}
$$

If z_3 and z_4 are zero, then $\dot{V}_1$ is negative and z_1 and z_2 will converge exponentially towards zero.

• *Step* 2: The derivative of z_3 and z_4 are expressed as

$$\dot{z}_3 = \beta_{1,1} u_1 + \beta_{1,2} u_2 + \boldsymbol{\phi}_1^\top \boldsymbol{\theta}_1 + c_1(x_3 - \dot{x}_{r1}) - \ddot{x}_{r1}, \tag{12.17}$$

$$\dot{z}_4 = -\beta_{2,1} u_1 + \beta_{2,2} u_2 + \boldsymbol{\phi}_2^\top \boldsymbol{\theta}_2 + c_2(x_4 - \dot{x}_{r2}) - \ddot{x}_{r2}. \tag{12.18}$$

The control inputs u_1 and u_2 will now be designed so that z_1, z_2, z_3, and z_4 all converge towards zero.

The adaptive control law is designed as follows:

$$\boldsymbol{u} = \begin{bmatrix} u_1 \\ u_2 \end{bmatrix} = \hat{\boldsymbol{B}}^{-1} \bar{\boldsymbol{u}} = \hat{\boldsymbol{R}} \bar{\boldsymbol{u}}, \tag{12.19}$$

where

$$\bar{\boldsymbol{u}} = \begin{bmatrix} \bar{u}_1 \\ \bar{u}_2 \end{bmatrix}, \quad \hat{\boldsymbol{B}} = \begin{bmatrix} \hat{\boldsymbol{\beta}}_1 \\ \hat{\boldsymbol{\beta}}_2 \end{bmatrix}, \tag{12.20}$$

$$\bar{u}_1 = -z_1 - \boldsymbol{\phi}_1^\top \hat{\boldsymbol{\theta}}_1 - c_3 z_3 - c_1(x_3 - \dot{x}_{r1}) + \ddot{x}_{r1}, \tag{12.21}$$

$$\bar{u}_2 = -z_2 - \boldsymbol{\phi}_2^\top \hat{\boldsymbol{\theta}}_2 - c_4 z_4 - c_2(x_4 - \dot{x}_{r2}) + \ddot{x}_{r2}, \tag{12.22}$$

$$\hat{\boldsymbol{\beta}}_1 = \begin{bmatrix} \hat{\beta}_{1,1} & \hat{\beta}_{1,2} \end{bmatrix}, \quad \hat{\boldsymbol{\beta}}_2 = \begin{bmatrix} -\hat{\beta}_{2,1} & \hat{\beta}_{2,2} \end{bmatrix}, \tag{12.23}$$

c_3 and c_4 are positive constants, $\hat{\theta}_1$, $\hat{\theta}_2$, $\hat{\beta}_{i,j}$ are the estimates of θ_1, θ_2, $\beta_{i,j}$ and $\hat{R}$ is the inverse of the matrix $\hat{B}$.

The parameter updating laws are chosen as

$$\dot{\hat{\theta}}_1 = \text{Proj}\{\Gamma_1 \phi_1 z_3\}, \tag{12.24}$$

$$\dot{\hat{\theta}}_2 = \text{Proj}\{\Gamma_2 \phi_2 z_4\}, \tag{12.25}$$

$$\dot{\hat{\beta}}_1^\top = \text{Proj}\{\Gamma_3 u z_3\}, \tag{12.26}$$

$$\dot{\hat{\beta}}_2^\top = \text{Proj}\{\Gamma_4 u z_4\}, \tag{12.27}$$

where Γ_k, $k \in \{1, 2, 3, 4\}$, are positive definite adaptation gain matrices and $\text{Proj}\{\cdot\}$ is the projection operator given in Appendix C, which ensures that the estimates and estimation errors are nonzero and within known bounds. Let $\tilde{\theta}_i = \theta_i - \hat{\theta}_i$ and $\tilde{\beta}_i = \beta_i - \hat{\beta}_i$, $i = 1, 2$, be the parameter estimation errors.

The projection operator $\dot{\hat{\theta}} = \text{Proj}\{\tau\}$ has the following property

$$-\tilde{\theta}^\top \Gamma^{-1} \text{Proj}\{\tau\} \le -\tilde{\theta}^\top \Gamma^{-1} \tau. \tag{12.28}$$

By using (12.19), we have

$$Bu = B\hat{R}\bar{u} = \bar{u} + \tilde{B}\hat{R}\bar{u} = \bar{u} + \tilde{B}u, \tag{12.29}$$

where $\tilde{B} = B - \hat{B}$. The determinant of matrix B will always be positive with the known signs of the parameters and from Assumption 12.2, where $\det(B) = \beta_{1,1}\beta_{2,2} + \beta_{2,1}\beta_{1,2}$, and from this and also with the projection operator, the matrix $\hat{R}$ does not have any singularities and is defined for all estimated parameters, given that the initial values are chosen positive. Now the terms $\beta_{1,1}u_1 + \beta_{1,2}u_2$ and $-\beta_{2,1}u_1 + \beta_{2,2}u_2$ in (12.17) and (12.18) can be expressed as

$$\beta_1 u = \bar{u}_1 + \tilde{\beta}_1 u, \tag{12.30}$$

$$\beta_2 u = \bar{u}_2 + \tilde{\beta}_2 u. \tag{12.31}$$

We define the final Lyapunov function as

$$\begin{aligned} V_2(z, \tilde{\beta}, \tilde{\theta}, t) = {} & V_1 + \frac{1}{2}z_3^2 + \frac{1}{2}z_4^2 + \frac{1}{2}\tilde{\theta}_1^\top \Gamma_1^{-1} \tilde{\theta}_1 \\ & + \frac{1}{2}\tilde{\theta}_2^\top \Gamma_2^{-1} \tilde{\theta}_2 + \frac{1}{2}\tilde{\beta}_1 \Gamma_3^{-1} \tilde{\beta}_1^\top + \frac{1}{2}\tilde{\beta}_2 \Gamma_4^{-1} \tilde{\beta}_2^\top. \end{aligned} \tag{12.32}$$

The derivative of (12.32) along with (12.17) to (12.31) gives

$$\dot{V}_2 = - c_1 z_1^2 - c_2 z_2^2 - c_3 z_3^2 - c_4 z_4^2 + \phi_1^\top \tilde{\theta}_1 z_3$$
$$+ \phi_2^\top \tilde{\theta}_2 z_4 - \tilde{\theta}_1^\top \Gamma_1^{-1} \dot{\hat{\theta}}_1 - \tilde{\theta}_2^\top \Gamma_2^{-1} \dot{\hat{\theta}}_2$$
$$+ \tilde{\beta}_1 u z_3 + \tilde{\beta}_2 u z_4 - \tilde{\beta}_1 \Gamma_3^{-1} \dot{\hat{\beta}}_1^\top - \tilde{\beta}_2 \Gamma_4^{-1} \dot{\hat{\beta}}_2^\top$$
$$= - c_1 z_1^2 - c_2 z_2^2 - c_3 z_3^2 - c_4 z_4^2$$
$$- \tilde{\theta}_1^\top \Gamma_1^{-1} \left(\dot{\hat{\theta}}_1 - \Gamma_1 \phi_1 z_3 \right) - \tilde{\beta}_1 \Gamma_3^{-1} \left(\dot{\hat{\beta}}_1^\top - \Gamma_3 u z_3 \right)$$
$$- \tilde{\theta}_2^\top \Gamma_2^{-1} \left(\dot{\hat{\theta}}_2 - \Gamma_2 \phi_2 z_4 \right) - \tilde{\beta}_2 \Gamma_4^{-1} \left(\dot{\hat{\beta}}_2^\top - \Gamma_4 u z_4 \right). \tag{12.33}$$

The property of the projection operator in (12.28) and the update laws (12.24)–(12.27) eliminate the last four terms in equation (12.33). Then

$$\dot{V}_2 \le - c_1 z_1^2 - c_2 z_2^2 - c_3 z_3^2 - c_4 z_4^2. \tag{12.34}$$

We then have the following stability and performance results based on the control scheme.

Theorem 12.1
Considering the closed-loop adaptive system consisting of the plant (12.3), the adaptive controller (12.19), the virtual control laws (12.13) and (12.14), the parameter updating laws (12.24)–(12.27) and Assumptions 12.1 and 12.2, all signals in the closed-loop system are ensured to be uniformly bounded. Furthermore, asymptotic tracking is achieved, i.e.

$$\lim_{t \to \infty} = [x_i(t) - x_{ri}(t)] = 0, \quad i = 1, 2. \tag{12.35}$$

Proof. The stability properties of the equilibrium follow from (12.32) and (12.34). By applying the LaSalle-Yoshizawa theorem, V_2 is uniformly bounded. This implies that z_1, z_2, z_3, z_4 are bounded and are asymptotically stable and z_1, z_2, z_3, $z_4 \to 0$ as $t \to \infty$ and also $\hat{\theta}_1$, $\hat{\theta}_2$, $\hat{\beta}_1$, and $\hat{\beta}_2$ are bounded. Since $z_1 = x_1 - x_{r1}$ and $z_2 = x_2 - x_{r2}$, tracking of the reference signals is also achieved, and x_1 and x_2 are also bounded since z_1 and z_2 are bounded and since x_{r1} and x_{r2} are bounded by definition, cf. Assumption 12.1. The virtual controls α_1 and α_2 are also bounded from (12.13) and (12.14) and then x_3 and x_4 are also bounded. From (12.19), it follows that the control inputs also are bounded.

Remark 12.1   Theorem 12.1 implies that the error signals will converge to zero. For a real system like the helicopter model, there are disturbances due to noise from sensors and unmodeled dynamics that are not included in this model, the helicopter will be stabilized with an adaptive controller, where the solution is ultimately bounded by a constant μ_0, that is, $\|z\| \le \mu_0$, $\forall t \ge T$, for some $T > 0$ [52].

12.3.2 Inputs Quantization

Considering the nonlinear state space model with quantized inputs expressed as

$$
\dot{\boldsymbol{x}} =
\begin{bmatrix}
x_3 \\
x_4 \\
\boldsymbol{\phi}_1^\top \boldsymbol{\theta}_1 \\
\boldsymbol{\phi}_2^\top \boldsymbol{\theta}_2
\end{bmatrix}
+
\begin{bmatrix}
0 \\
0 \\
\beta_{1,1} q(u_1) + \beta_{1,2} q(u_2) \\
-\beta_{2,1} q(u_1) + \beta_{2,2} q(u_2)
\end{bmatrix},
\tag{12.36}
$$

where the control inputs u_1 and u_2 are quantized by the uniform quantizer defined in (12.8). The change of coordinates and Step 1 will be the same as when the inputs are continuous and the virtual control laws are designed as in (12.13) and (12.14). In step 2 the control inputs appear, and the derivative of z_3 and z_4 are expressed as

$$
\dot{z}_3 = \beta_{1,1} q(u_1) + \beta_{1,2} q(u_2) + \boldsymbol{\phi}_1^\top \boldsymbol{\theta}_1 + c_1(x_3 - \dot{x}_{r1}) - \ddot{x}_{r1},
\tag{12.37}
$$

$$
\dot{z}_4 = -\beta_{2,1} q(u_1) + \beta_{2,2} q(u_2) + \boldsymbol{\phi}_2^\top \boldsymbol{\theta}_2 + c_2(x_4 - \dot{x}_{r2}) - \ddot{x}_{r2}.
\tag{12.38}
$$

The quantizer inputs are decomposed into two parts

$$
q(u_k) = u_k(t) + d_k(t),
\tag{12.39}
$$

where d_k is the quantization error and bounded by a constant, $|d_k| \le \delta_k$, where

$$
\delta_k = \max\{u_{k,0} + l_k/2, l_k/2\}.
\tag{12.40}
$$

Thus the equations (12.37) and (12.38) are expressed as

$$
\begin{aligned}
\dot{z}_3 =&\, \beta_{1,1}(u_1 + d_1) + \beta_{1,2}(u_2 + d_2) + \boldsymbol{\phi}_1^\top \boldsymbol{\theta}_1 \\
&+ c_1(x_3 - \dot{x}_{r1}) - \ddot{x}_{r1},
\end{aligned}
\tag{12.41}
$$

$$
\begin{aligned}
\dot{z}_4 =&\, -\beta_{2,1}(u_1 + d_1) + \beta_{2,2}(u_2 + d_2) + \boldsymbol{\phi}_2^\top \boldsymbol{\theta}_2 \\
&+ c_2(x_4 - \dot{x}_{r2}) - \ddot{x}_{r2},
\end{aligned}
\tag{12.42}
$$

where due to quantization, two extra terms are included in each equation. The inputs are designed in the controller (12.19) together with (12.21)–(12.23) and with the parameter updating laws (12.24)–(12.27). The final Lyapunov function V_2 is defined as in (12.32), the same as without quantization. Then the derivative of V_2 gives

$$
\begin{aligned}
\dot{V}_2 =&\, -c_1 z_1^2 - c_2 z_2^2 - c_3 z_3^2 - c_4 z_4^2 + \beta_1 d z_3 + \beta_2 d z_4 \\
&- \tilde{\boldsymbol{\theta}}_1^\top \boldsymbol{\Gamma}_1^{-1} \left(\dot{\hat{\boldsymbol{\theta}}}_1 - \boldsymbol{\Gamma}_1 \boldsymbol{\phi}_1 z_3 \right) - \tilde{\beta}_1 \boldsymbol{\Gamma}_3^{-1} \left(\dot{\hat{\beta}}_1 - \boldsymbol{\Gamma}_3 u z_3 \right) \\
&- \tilde{\boldsymbol{\theta}}_2^\top \boldsymbol{\Gamma}_2^{-1} \left(\dot{\hat{\boldsymbol{\theta}}}_2 - \boldsymbol{\Gamma}_2 \boldsymbol{\phi}_2 z_4 \right) - \tilde{\beta}_2 \boldsymbol{\Gamma}_4^{-1} \left(\dot{\hat{\beta}}_2 - \boldsymbol{\Gamma}_4 u z_4 \right),
\end{aligned}
\tag{12.43}
$$

where $d = [d_1 \ d_2]^\top$, and the property of the projection operator in (12.28) and the update laws (12.24)–(12.27) eliminate the last four terms in equation (12.43). Then

$$
\begin{aligned}
\dot{V}_2 \leq & -c_1 z_1^2 - c_2 z_2^2 - c_3 z_3^2 - c_4 z_4^2 + \beta_1 d z_3 + \beta_2 d z_4 \\
\leq & -c_0 \|z\|^2 + \sqrt{(|\beta_1|\delta)^2 + (|\beta_2|\delta)^2} \|z\| \\
\leq & -(1-\lambda) c_0 \|z\|^2 - \lambda c_0 \|z\|^2 + \sqrt{(|\beta_1|\delta)^2 + (|\beta_2|\delta)^2} \|z\| \\
\leq & -(1-\lambda) c_0 \|z\|^2, \quad \forall \|z\| \geq \frac{\sqrt{(|\beta_1|\delta)^2 + (|\beta_2|\delta)^2}}{\lambda c_0}
\end{aligned}
\tag{12.44}
$$

where $c_0 = \min\{c_1, c_2, c_3, c_4\}$, the constant $\delta = [\delta_1 \ \delta_2]^\top$ is the maximum quantization errors as defined in (12.40) and $0 < \lambda < 1$. We then have the following stability and performance results based on the control scheme.

Theorem 12.2
Consider the closed-loop adaptive system consisting of the plant (12.36), the adaptive controller (12.19), the virtual control laws (12.13) and (12.14), the parameter updating laws (12.24)–(12.27), the uniform quantizer (12.8) and Assumptions 12.1 and 12.2, all signals in the closed-loop system are ensured to be uniformly bounded. The tracking error will converge to a compact set, i.e.

$$
\|z\| \leq \mu = \frac{\sqrt{(|\beta_1|\delta)^2 + (|\beta_2|\delta)^2}}{\lambda c_0},
\tag{12.45}
$$

where μ is a positive constant. The tracking errors $e_i(t) = x_i(t) - x_{ri}(t)$ are ultimately bounded by $\|e_i\| \leq \mu$, and tracking is achieved.

Proof. The stability properties of the equilibrium follows from (12.32) and (12.44). The quantization error is bounded by definition (12.40). By applying the LaSalle-Yoshizawa theorem, V_2 is bounded. This implies that $z_1, z_2, z_3, z_4, \hat{\theta}_1, \hat{\theta}_2, \hat{\beta}_1$, and $\hat{\beta}_2$ are bounded. Furthermore, z_1, z_2, z_3, and z_4, will converge to a compact set containing the equilibrium as $t \to \infty$. Since $z_1 = x_1 - x_{r1}$ and $z_2 = x_2 - x_{r2}$, the states x_1 and x_2 are also bounded since z_1 and z_2 are bounded and since x_{r1} and x_{r2} are bounded by definition, cf. Assumption 12.1. Tracking of the reference signals is achieved, with a bounded tracking error. The virtual controls α_1 and α_2 are also bounded from (12.13) and (12.14) and then x_3 and x_4 are also bounded. From (12.19), it follows that the control inputs also are bounded.

Remark 12.2 The tracking errors are adjustable by tuning the design parameters c_i, $i \in \{1, 2, 3, 4\}$. The smaller quantization intervals l_k, the smaller the compact set for the error variables $\|z\|$ will be, and if l_k decreases to zero and there is no quantization, the error will also be zero and the result will be similar to Theorem 12.1, without quantization. The bound for the error system will also include the bound from Remark 12.1 for the helicopter model, only shifting the bound to $\|z\| \leq \mu_0 + \mu$, $\forall t \geq T$, for some $T > 0$.

12.4   Experimental Results

The Quanser Aero helicopter system shown in Figure 12.3 is a two-rotor laboratory equipment for flight control-based experiments. The setup is a horizontal position of the main thruster and a vertical position of the tail thruster, which resembles a helicopter with two propellers driven by two DC motors.

Figure 12.3 Quanser Aero

The proposed controller was simulated using MATLAB/Simulink and tested on the Quanser Aero helicopter system. The initial states were set as $x(0) = 0$ and the design parameters were set as $c_1 = c_2 = 6$, $c_3 = c_4 = 3$, $\Gamma_1 = I_3$, $\Gamma_2 = I_2$, and $\Gamma_3 = \Gamma_4 = 0.01I_2$. The same quantization intervals were used for the two inputs, since the two motors on the helicopter model equal and the range of their inputs are $[-24, 24]$. The interval was chosen $l_1 = l_2 = 1$, and is a quantization level to show the effect of the quantization, since there are other disturbances that will affect the results as e.g. noise from sensors. The constant $u_{k,0}$ was chosen as 0 for both inputs, and so the upper bound for the quantization errors were $\delta_1 = \delta_2 = l_k/2 = 1/2$. The initial values for the parameters were chosen as $\hat{\beta}_1(0) = [0.0506 \quad 0.0506]$, $\hat{\beta}_2(0) = [-0.0645 \quad 0.0810]$, $\hat{\theta}_1(0) = [0.322 \quad 1.8436 \quad 0.0007]^\top$, and $\hat{\theta}_2(0) = [0.4374 \quad 0.0014]^\top$.

The objective of this test is to make the pitch angle track a reference and the rotation of the yaw is zero. The reference signal is a sinusoidal signal with an amplitude of 40 degrees and a frequency of 0.05 Hz.

12.4.1 Results without Quantization

The results from simulation and experiment on the Aero helicopter with continuous inputs are shown in Figures 12.4–12.5, where red plots are from simulation and blue plots are from the experiment. While the results in Figure 12.4 show that the tracking error converges to zero in the simulation, while the tracking error converges to a compact set in the experiment. This is due to the unknown disturbances affecting the system as in Remark 12.1. It is shown that the tracking of the reference signals is achieved for both pitch and yaw angles.

In Figure 12.5, the norm of z is plotted. The results shows that tracking error $\|z\| \to 0$ as $t \to \infty$ in the simulation, while the tracking error is within a compact set in the experiment, where the bound $\mu_0 = \max\|z\|$.

12.4.2 Results with Quantization

When the inputs were quantized, the results are plotted in Figures 12.6–12.8. From Figure 12.6, it can be seen that the pitch followed the desired trajectory of a sine wave using the proposed adaptive controller both in simulation and experiment on the helicopter system. In Figure 12.7, the norm of the tracking error z is plotted and the bound μ is ploted whcih is computed with $\lambda = 0.999$, $\beta_1 = \hat{\beta}_1(0)$, and $\beta_2 = \hat{\beta}_2(0)$, as given in Theorem 12.2. In the transient period, the norm of the error is within a bound. Figure 12.8 shows that $\|z\|$ is within the bound for the whole time period, where μ_0 is the bound calculated for the system without quantization and $\mu_0 + \mu$ is the bounded for the system with quantization.

12.4.3 Comparing Results

To compare the results with and without quantization, the total tracking error z_{track} and the total voltage u_{total} was measured, where

$$z_{track} = \sum_{i=1}^{2} \int_0^t z_i(\tau)^2 d\tau, \tag{12.46}$$

$$u_{total} = \sum_{i=1}^{2} \int_0^t u_i(\tau)^2 d\tau, \tag{12.47}$$

with $t = 50\ s$. There is a trade-off between the error and voltage consumption since the more accurate the controller is, the more voltage is needed to hold the trajectory closer to the reference.

In Table 12.1, the results are compared for different quantization intervals. The tracking error is higher when the inputs are quantized, while the total voltage is lower

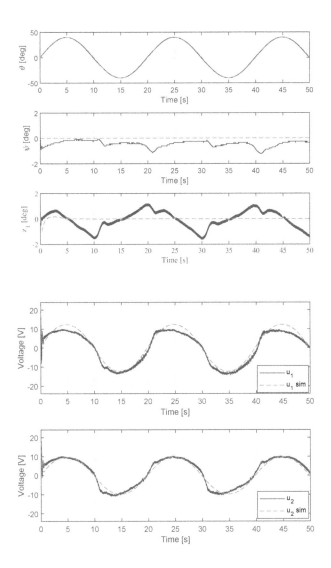

Figure 12.4 Results without quantization: 1) pitch angle, 2) yaw angle, 3) pitch angle error, 4) inputs

for most of the tests with quantization. The higher error is due to the quantization error, as expected from Theorems 12.1 and 12.2.

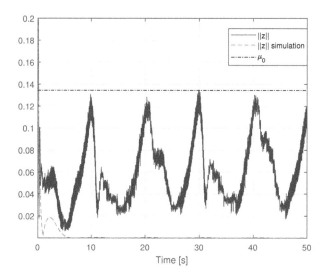

Figure 12.5 Norm of z without quantization

Table 12.1   Comparison of Error and Voltage Use with and without Quantization

Measurement	Quantization				
	$l_k = 0$	$l_k = 0.1$	$l_k = 0.5$	$l_k = 1$	$l_k = 1.5$
z_{track}	0.0110	0.0116	0.0114	0.0116	0.0121
u_{total}	6429	6444	6365	6367	6328

12.5   Notes

In this chapter, an adaptive backstepping control scheme is developed for a MIMO nonlinear helicopter model with input quantization. The system parameters are not required to be fully known for the controller design. A theoretical proof of stability is given with the use of constructed Lyapunov functions, where boundedness of all signals in the closed-loop system are achieved and the tracking error converges to a compact set.

Acknowledgment

Reprinted from Copyright (2021), with permission from Elsevier. Siri Marte Schlanbush, Jing Zhou, "Adaptive Backstepping Control of a 2-DOF Helicopter System with Uniform Quantized Inputs", *The 46th Annual Conference of the IEEE Industrial Electronics Society*, pp. 88–94, 2020.

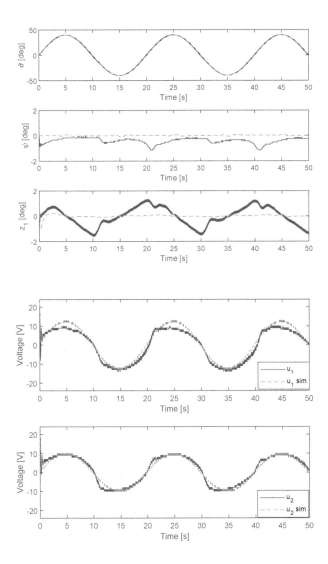

Figure 12.6 Results with quantization: 1) pitch angle, 2) yaw angle, 3) pitch angle error, 4) inputs

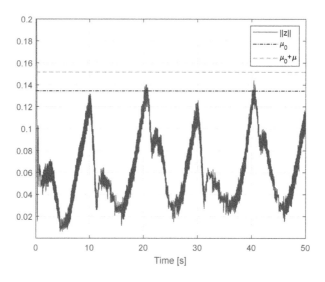

Figure 12.7 Simulation of norm of z with quantization and the bound μ

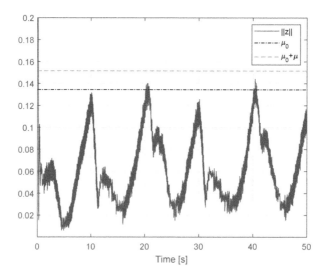

Figure 12.8 Norm of z with quantization from the helicopter, with bounds

Chapter 13

Quantized Distributed Secondary Control for DC Microgrid

The control of DC microgrid is becoming increasingly important in modern power systems. One important control objective for such a system is to ensure DC bus voltage stability and precise current sharing among converters. This chapter presents a quantized distributed secondary control strategy for current sharing and voltage restoration in DC microgrid. Through this strategy, each converter can only transmit the quantized signals to its neighbors. Therefore, the communication burden among converters is reduced. Moreover, the proposed strategy also enables the DC microgrid to connect various loads, including both linear and non-linear loads.

13.1   Introduction

Microgrid has been gaining considerable attention for its flexibility in integrating various types of renewable energy sources (RESs) [55, 123]. The recent literature on this topic mainly focuses on AC microgrid as the utility grid depends on AC systems [10,25,77]. However, a number of different RESs, such as solar energy, naturally produces direct current (DC) power. Therefore, a DC micrgrid consisting of DC components is more convenient for control. In addtion, DC source generates no reactive power, thus it is also able to supply power in a more reliable and efficient way [9, 32].

DOI: 10.1201/9781003176626-13   **195**

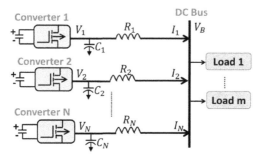

Figure 13.1 The DC system setup

Proper current sharing among DC sources is essential in DC microgrid as it is the way to achieve power dispatch on loads [68, 69]. In islanded DC microgrid, droop control is one of the principle control methods to achieve precise current sharing [4, 31]. The key idea of droop control is to set a sufficiently large droop gain to dominate the line resistance. With this gain determined, droop control is able to ensure desired current sharing among all the DC sources involved. However, a large droop gain could lead the bus voltage to deviate severely from its reference value, which is unacceptable for safe grid operation. Therefore, the limitations of droop control have to be addressed to achieve accurate current sharing and voltage stability at the same time.

To overcome the limitations of droop control, another control layer, i.e. secondary control, is proposed in [30]. With the help of secondary control, proper bus voltage regulation can be achieved with the desired current sharing property unaffected. Since then, tremendous efforts have been made in distributed secondary control in the existing literature, see [1, 22, 25, 31, 78, 122, 151] and reference therein for examples. However, so far distributed secondary control results with quantized signal transmissions are still very limited. In this chapter, based on the concept "virtual voltage drop" proposed in [122], a quantized distributed secondary control strategy is presented. This strategy can simultaneously guarantee both proper current sharing and DC bus voltage regulation. In addition, it also enables the DC microgrid to support various types of linear and nonlinear loads.

13.2 Problem Formulation and Control Objective

Similar to [31, 68], a DC microgrid with a bus connected with N converters is considered in this chapter, as shown in Figure 13.1. It is worth pointing out that this DC system is widely used in different systems such as the More Electric Aircraft system. In this section, the limitations of traditional droop control will be first discussed. Then, the control objective will be given accordingly.

13.2.1 Droop Control

In traditional droop control, a droop function is designed to set the reference voltage value V_i^{ref} for each converter. Specifically, for the ith converter ($i = 1, \ldots, N$), the droop function is given as below

$$V_i^{ref} = V^* - k_i I_i \tag{13.1}$$

where V^* is the desired value for the bus voltage, k_i is the droop gain, and I_i is the output current of converter i. For each converter, two PI controllers located in the voltage and current control loop in the primary control layer are used. By properly tuning the PI parameters, the output voltage V_i of converter i can track V_i^{ref} with a very fast speed [31], thus it is safe to have $V_i = V_i^{ref}$.

Let R_i be the resistance of the feeder connecting converter i to the common DC bus. Then, the bus voltage V_B is

$$V_B = V_i - R_i I_i = V^* - (R_i + k_i) I_i \tag{13.2}$$

From (13.2), it is obtained that

$$(R_i + k_i) I_i = (R_j + k_j) I_j, \quad \forall i, j = 1, \ldots, N \tag{13.3}$$

From (13.3) we further have

$$I_i / I_j = (R_j + k_j) / (R_i + k_i) \tag{13.4}$$

Since k_i is always set considerably larger than R_i, i.e. $k_i \gg R_i$, we have $I_i / I_j = k_j / k_i$. Therefore, by setting the droop gain large enough, proper current sharing is achieved.

13.2.2 Control Objective

As discussed above, the control gain k_i should be set considerably large to ensure proper current sharing. However, from (13.2), it is observed that a large k_i will increase the voltage drop and deviate the bus voltage V_B severely from its nominal voltage V^*. Therefore, a secondary control layer is needed to restore the DC bus voltage back to its nominal value. Specifically, the control objective is to design a quantized distributed secondary control strategy which can guarantee current sharing and voltage restoration at the same time, i.e.

$$\lim_{t \to \infty} I_i / I_j = k_j / k_i \tag{13.5}$$

$$\lim_{t \to \infty} e = V_B - V^* = 0 \tag{13.6}$$

13.2.3 Data Communication Network

A communication network among the converters is needed for the development of distributed control strategy. Consider the graph $\mathcal{G} = (\mathcal{V}, \mathcal{E}, \mathcal{A})$, in which $\mathcal{V}$ is the set containing all DC converters, $\mathcal{E} \subseteq \mathcal{V} \times \mathcal{V}$ denotes the edge set, and $\mathcal{A} \in \mathbb{R}^{N \times N}$ is the adjacency matrix. Denoting a_{ij} as the entities of $\mathcal{A}$, then $a_{ij} > 0$ if $(v_j, v_i) \in \mathcal{E}$, and $a_{ij} = 0$ if $(v_j, v_i) \notin \mathcal{E}$. The Laplacian matrix L of graph $\mathcal{G}$ is made up by $L_{ii} = \sum_{i \neq j} a_{ij}$, and $L_{ij} = -a_{ij}$ for $i \neq j$. Define $Sym(L) = \frac{1}{2}(L + L^T)$, then if $\mathcal{G}$ is strongly connected and weight-balanced, zero is a single eigenvalue of both L and $Sym(L)$. Let $\hat{\lambda}_1, \ldots, \hat{\lambda}_N$ denote the eigenvalues of $Sym(L)$, we define $0 = \hat{\lambda}_1 < \hat{\lambda}_2 \leq \hat{\lambda}_3 \leq \ldots \leq \hat{\lambda}_N$, where $\hat{\lambda}_i$ is the ith eigenvalue.

It is assumed that the converter communication graph is strongly connected and weight-balanced. In practice, the communication link is always bidirectional, thus the required communication topology is naturally met when the involved converters are connected.

13.3 Quantized Distributed Secondary Control

This section will present a distributed secondary control strategy with quantized signal transmissions to achieve (13.5)–(13.6). Since the designed secondary control signal u_i is added to the droop function (13.1), we have

$$V_i^{ref} = V^* - k_i I_i + u_i \tag{13.7}$$

Then, we obtain

$$V_B = V^* - (k_i + R_i)I_i + u_i \tag{13.8}$$

Define a parameter V_i^d as

$$V_i^d = (k_i + R_i)I_i \tag{13.9}$$

which is time-varying. It is shown in [122] that V_i^d represents a parameter describing the "virtual voltage drop". Then, the distributed secondary controller is designed as

$$\begin{cases} \dot{x}_i = -\alpha x_i - \beta \Sigma_{j \in N_i} a_{ij}(\hat{u}_i - \hat{u}_j) \\ u_i = x_i + V_i^d, \quad i = 1, \ldots, N \end{cases} \tag{13.10}$$

where α and β are positive constants, and $\hat{u}_i = q_u(u_i)$ is the quantized signal of u_i by the hysteresis-uniform quantizer defined in (3.3).

The whole structure including the primary and secondary control layer for converter i is shown in Figure 13.2. As can be seen, only the quantized signal of u_i, i.e. $\hat{u}_i$, needs to be transmitted to its neighbors.

Theorem 13.1
With the designed controller (13.10), if the converter communication graph is strongly connected and weight-balanced, the control objective (13.5)–(13.6) can be ensured.

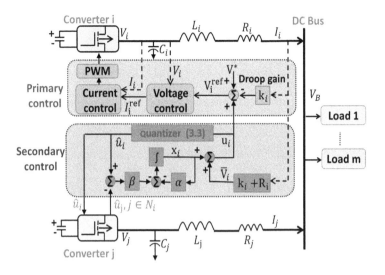

Figure 13.2 Quantized distributed secondary control

Proof 13.1 From (13.8), we have

$$e = u_i - V_i^d, \ i = 1, \dots, N \tag{13.11}$$

Summing (13.11) together gives

$$Ne = \sum_{i=1}^{N} u_i - \sum_{i=1}^{N} V_i^d = \sum_{i=1}^{N}(u_i - V_{avg}^d) = \sum_{i=1}^{N} \tilde{e}_i \tag{13.12}$$

in which $V_{avg}^d = \frac{1}{N}\sum_{i=1}^{N} V_i^d$ and $\tilde{e}_i = u_i - V_{avg}^d$. Define the quantization error as $e_i = \hat{u}_i - u_i$, then based on (13.10) and (13.11) we obtain

$$\begin{aligned} \dot{\tilde{e}}_i =& \dot{V}_i^d - \dot{V}_{avg}^d - \alpha(u_i - V_i^d) - \beta\Sigma_{j\in N_i}a_{ij}(\tilde{e}_i - \tilde{e}_j + e_i - e_j) \\ =& -\alpha(u_i - V_{avg}^d) + \dot{V}_i^d - \dot{V}_{avg}^d + \alpha V_i^d - \alpha V_{avg}^d \\ & -\beta\Sigma_{j\in N_i}a_{ij}(\tilde{e}_i - \tilde{e}_j) - \beta\Sigma_{j\in N_i}a_{ij}(e_i - e_j) \end{aligned} \tag{13.13}$$

By defining $\tilde{e} = [\tilde{e}_1, \dots, \tilde{e}_N]^T$, $V^d = [V_1^d, \dots, V_N^d]^T$, $\bar{e} = [e_1, \dots, e_N]^T$, and $\Pi_N = \mathbf{I}_N - \frac{1}{N}\mathbf{1}_N\mathbf{1}_N^T$, we have

$$\dot{\tilde{e}} = -\alpha\tilde{e} - \beta L\tilde{e} + \Pi_N(\dot{V}_i^d + \alpha V_i^d) - \beta L\bar{e} \tag{13.14}$$

Next, we take the following transformation

$$\tilde{e}_T = \begin{bmatrix} \tilde{e}_{T1} \\ \tilde{e}_{T2} \end{bmatrix} = \underbrace{\begin{bmatrix} \chi^T \\ S^T \end{bmatrix}}_{B} \tilde{e} \tag{13.15}$$

where $\chi = \frac{1}{\sqrt{N}}\mathbf{1}_{N\times1}$, S is a matrix rendering $\left[\frac{1}{\sqrt{N}}\mathbf{1}_{N\times1} \quad S\right]$ orthonormal. Moreover, $S^T S = \mathbf{I}_{N-1}$, $SS^T = \Pi_N$ and $BB^T = \mathbf{I}_N$.

Since $\chi^T L = 0$ and $\chi^T \Pi_N = 0$, (13.14)–(13.15) gives $\dot{\tilde{e}}_{T1} = -\alpha\tilde{e}_{T1}$ and

$$\dot{\tilde{e}}_{T2} = -(\beta S^T LS + \alpha I_{N-1})\tilde{e}_{T2} + S^T \Pi_N(\dot{V}_i^d + \alpha V_i^d) \\ - \beta S^T L\bar{e} \tag{13.16}$$

Thus, we have that $|\tilde{e}_{T1}| \leq |\tilde{e}_{T1}(0)|e^{-\alpha t}$ and

$$\tilde{e}_{T2} = \Phi(t,0)\tilde{e}_{T2}(0) + \int_0^t \Phi(t,s)S^T\Pi_N(\dot{V}_i^d + \alpha V_i^d)ds \\ - \int_0^t \beta\Phi(t,s)S^T L\bar{e}ds \tag{13.17}$$

in which $\Phi(t,s) = e^{-(\beta S^T LS + \alpha I_{N-1})(t-s)}$. Let $A = -(\beta S^T LS + \alpha I_{N-1})$ and λ_A be the eigenvalue of $\frac{1}{2}(A + A^T)$. Then,

$$|\lambda_A I_{N-1} - \frac{1}{2}(A + A^T)| \\ = |(\lambda_A + \alpha)I_{N-1} + \beta S^T LS| = 0 \tag{13.18}$$

As the eigenvalues of $S^T LS$ are $\hat{\lambda}_i, i = 2,\ldots,N$, $\lambda_A = -\alpha - \beta\hat{\lambda}_i$, based on [6], we obtain

$$\Phi(t,s) \leq e^{(-\alpha-\beta\hat{\lambda}_2)(t-s)} \tag{13.19}$$

As $\dot{V}_i^d$, V_i^d and e_i are all bounded, we have $||S^T\Pi_N(\dot{V}_i^d + \alpha V_i^d)|| \leq ||\Pi_N(\dot{V}_i^d + \alpha V_i^d)|| = \gamma$ and $||S^T L\bar{e}|| \leq ||L|| \, ||\Delta||$, where γ is a positive constant, and Δ denotes the upper bound of quantization errors for the quantizers used. As a result, we can get

$$||\tilde{e}_{T2}(t)|| \leq e^{(-\alpha-\beta\hat{\lambda}_2)t}||\tilde{e}_{T2}(0)|| \\ + \frac{\gamma + \beta||L|| \, ||\Delta||}{\beta\hat{\lambda}_2}(1 - e^{(-\alpha-\beta\hat{\lambda}_2)t})$$

and thus

$$\lim_{t\to\infty} ||\tilde{e}_{T2}(t)|| \leq \frac{\gamma + \beta||L|| \, ||\Delta||}{\beta\hat{\lambda}_2} \tag{13.20}$$

Since $||\tilde{e}|| = \tilde{e}_T^T BB^T \tilde{e}_T^T = ||\tilde{e}_T||$, we obtain

$$\lim_{t\to\infty} ||\tilde{e}|| \leq \lim_{t\to\infty}(\tilde{e}_{T1} + ||\tilde{e}_{T2}||) \leq \frac{\gamma + \beta||L|| \, ||\Delta||}{\beta\hat{\lambda}_2} \tag{13.21}$$

From (13.21), it is obtained that through increasing β and decreasing Δ, the bound of $||\tilde{e}||$ can be made arbitrarily small, i.e.,

$$\lim_{t\to\infty} |\tilde{e}_i| \leq \lim_{t\to\infty} ||\tilde{e}|| \approx 0, \quad i = 1, \ldots, N \tag{13.22}$$

Therefore, from (13.11), (13.12), and (13.22), we have $\lim_{t\to\infty} e \approx 0$. As a result, the control objective (13.6) is achieved. Further, it is easy to get that

$$\lim_{t\to\infty} V_i^d = \lim_{t\to\infty} V_j^d = V_{avg}^d, \quad \forall i, j = 1, \ldots, N \tag{13.23}$$

Since $k_i \gg R_i$, from (13.9) and (13.23), we have $I_i/I_j = k_j/k_i$. Therefore, (13.5) is ensured.

13.4   Case Studies

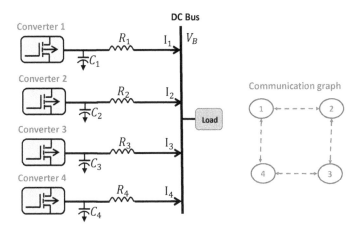

Figure 13.3 The case study system

As shown in Figure 13.3, a DC microgrid with four converters is built in the MATLAB/Simulink environment. Based on this system, two case studies are provided to demonstrate the effectiveness of the proposed quantized control strategy. Specifically, the first case tests linear resistive loads, while the second one examines nonlinear ZIP loads, i.e. constant impedance loads (Z), constant current loads (I), and constant power loads (P). In this system setup, two PI controllers are adopted in the primary control layer. The reference for the bus voltage is $V^* = 48V$. Following [31], the other parameters for the microgrid system are given in Table 13.1.

Table 13.1 System Setup [31]

Parameters	Descriptions	Values
V_{DC}	DC-link voltage	100 V
$C_i, i = 1, 2, 3, 4$	Filter capacitor	2200 μF
L_f	Filter inductance	20 mH
K_{VP}, K_{VI}	PI in voltage loop	4, 800
K_{IP}, K_{II}	PI in current loop	5, 110
f_{sw}	Switching frequency	20 kHz
$R_1, R3$	Line resistance	0.02 Ω
R_2, R_4	Line resistance	0.01 Ω
k_1	Droop gain	6
k_2	Droop gain	3
k_3, k_4	Droop gain	2, 2

13.4.1   Case 1: Resistive Load

This case examines the proposed strategy with resistive loads. The parameters of controller (13.10) are set as $\alpha = 20$, $\beta = 5$. For simplicity, the same hysteresis-uniform quantizer is employed for the four converters, and its parameters are set as $h = 0.05$ and $l = 0.1$ (please see the parameter definitions in Equation (3.3)). In this case study, the microgrid undergoes the following four steps:

Step 1 (0–2s): A load 5Ω is connected and only droop control is implemented;
Step 2 (2–6s): The presented secondary controller is activated at $t = 2s$;
Step 3 (6–10s): Another 5Ω is added at $t = 6s$;
Step 4 (10–14s): One 5Ω load is disconnected at $t = 10s$;

The output current and voltage of the four converters are shown in Figures.13.4–13.5, while the control signal u_2 and its quantized value $\hat{u}_2$ are presented in Figure 13.6. The other three control signals are omitted here for their similarity to u_2. As can be observed from Figures 13.4–13.5, if only droop control is implemented, though the current is properly shared, the bus voltage stays at around 42V instead of the nominal value 48V. When the presented secondary controller is implemented, the bus voltage quickly restores to 48V. Even with load changes, the voltage stability remains satisfactory. Moreover, proper current sharing is also well maintained as $I_1 : I_2 : I_3 : I_4 = \frac{1}{k_1} : \frac{1}{k_2} : \frac{1}{k_3} : \frac{1}{k_4}$ as desired. This case study shows the effectiveness of the quantized secondary control strategy proposed in this chapter.

13.4.2   Case 2: ZIP Loads

Similar to the work in [79], the control performance of the proposed strategy on ZIP loads is examined in this case. The control parameters remain the same as those in Case 1, and the system undergoes four steps below:

Step 1 (0–2s): A 400W constant power load is connected and only droop control is in effect;

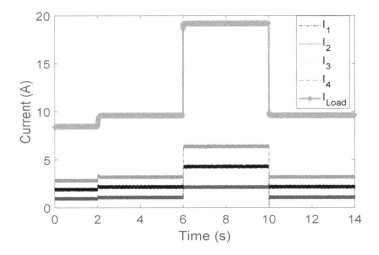

Figure 13.4 Current output

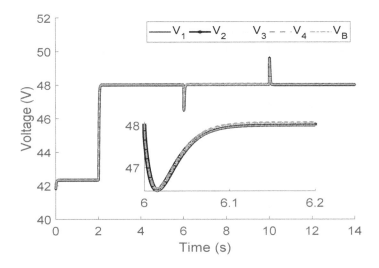

Figure 13.5 Voltage output

Step 2 $(2-6s)$: The proposed strategy is activated at $t = 2s$;
Step 3 $(6-10s)$: A constant impedance load 10Ω is added at $t = 6s$;
Step 4 $(10-14s)$: A constant current load 5A is added at $t = 10s$;

Figures 13.7–13.8 respectively present the output current and voltage of the four converters, and Figure 13.9 shows the secondary control signal u_2 and its quantized value $\hat{u}_2$. As observed from Figures 13.7–13.8, only with droop control, the bus

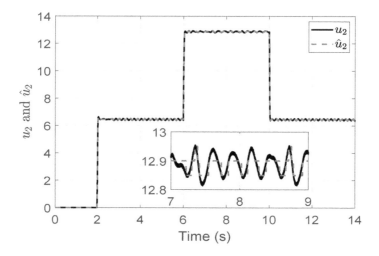

Figure 13.6 Control signal u_2 **and** $\hat{u}_2$

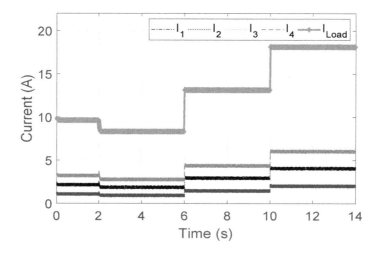

Figure 13.7 Current output

voltage cannot stay at the desired 48V. However, with the proposed secondary control strategy, both voltage regulation and accurate current sharing are well guaranteed, even with different types of load changes. This case demonstrates that the proposed quantized control strategy is able to handle nonlinear ZIP loads.

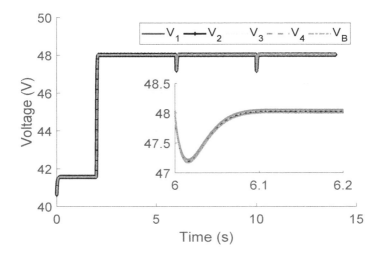

Figure 13.8 Voltage output

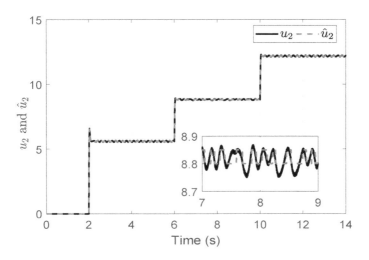

Figure 13.9 Control signal u_2 and $\hat{u}_2$

13.5   Notes

In this chapter, a quantized distributed secondary control strategy for DC microgrid is proposed. With the help of quantizers, each converter only needs to send the quantized signals to its neighbors. It is proved that the proposed strategy can ensure proper current sharing and DC bus voltage regulation at the same time. Moreover, it also enables the DC microgird to connect various kinds of loads. The effectiveness of the proposed strategy is also demonstrated by two case studies.

Chapter 14

Conclusions and Future Challenges

14.1 Conclusions

This book aims at presenting innovative technologies for designing and analyzing adaptive backstepping control systems involving quantized signals. The main control objectives are to achieve desired regulation of the system outputs while ensuring the boundedness of all closed-loop signals. Compared with the existing literature in related areas, novel solutions to the following challenging problems are provided in series.

■ Design of suitable adaptive backstepping control schemes for uncertain non-linear system with quantization.

■ Development of a systematic stability analysis for quantized control systems.

■ Relaxation of the stability conditions for adaptive control of uncertain nonlinear systems with input quantization (Chapters 4, 5, 6, 7);

■ Adaptive compensation for input quantization (Chapters 4, 5, 6, 7);

■ Adaptive compensation for state quantization (Chapters 8, 9);

■ Adaptive compensation for input and state quantization (Chapters 10, 11);

■ Adaptive compensation for input and output quantization (Chapter 12);

■ Decentralized adaptive control of uncertain interconnected systems with input quantization (Chapter 6);

DOI: 10.1201/9781003176626-14

■ State-feedback vs. Output-feedback

Chapters 4–11 can be further classified as state-feedback control (Chapters 4–6, 8–10) and output-feedback control (Chapters 7, 11). As full state measurement is absent in the latter class of control problems, observers are often needed to provide state estimates with sufficient accuracy. In [53], filters are developed to construct the state estimate, with which the estimation error can converge to zero exponentially if the observer is implementable with known system parameters. Based on this, the standard filters in [53] are adopted in Chapter 7 to estimate the local state variables. However, since the effects of input quantization are encompassed, the dynamics of the achieved state estimation error changes. This results in a more complicated process in adaptive control design and stability analysis. In Chapter 11, modified adaptive state observers are designed to estimate the local state variables. Since the effects of both input and output quantization are encompassed, the dynamics of the achieved state estimation error becomes complicated. By using transformations of coordinates, the proposed scheme ensures the global boundedness of all the states of the uncertain system.

■ Tracking Error Performances

In Chapters 4–12, the ultimate bound of tracking error is shown to be bounded by functions of design parameters including the control parameters and quantization parameters. This implies that the tracking error performances can be adjusted by suitably choosing these parameters. In fact, providing a promising way to improve the tracking performance of adaptive control systems by appropriately tuning design parameters is one of the prominent advantages of adaptive backstepping control over the conventional approaches, as stated in Chapter 1 and references therein.

14.2 Future Challenges

Some interesting problems which have not been extensively explored in decentralized/distributed adaptive control of interconnected systems and related areas are suggested as follows.

■ Extension of the developed control methodologies to time-varying systems.

■ Decentralized adaptive compensation for interconnected systems with state quantization, input and state quantization, and input and output quantization.

■ Distributed adaptive consensus/formation/containment control of uncertain multi-agent systems with quantized signals.

■ Distributed adaptive control of uncertain networked systems with quantization and event-triggered communication.

- Distributed adaptive resilient control of networked systems with malicious cyber attacks under state constraints.

- Application of adaptive quantized control to practical systems, such as robotic systems, offshore crane systems, drilling systems, formation of unmanned aerial vehicles (UAVs), etc.

Appendix A

Lyapunov Stability [53]

For all control systems and adaptive control systems in particular, stability is the primary requirement. Consider the time-varying system

$$\dot{x} = f(x, t) \tag{A.1}$$

where $x \in R^n$, and $f : \mathscr{R}^n \times \mathscr{R}_+ \to \mathscr{R}^n$ is piecewise continuous in t and locally Lipschiz in x. The solution of (A.1) which starts from the point x_0 at time $t_0 \geq 0$ is denoted as $x(t; x_0, t_0)$ with $x(t_0; x_0, t_0) = x_0$. If the initial condition x_0 is perturbed to $\tilde{x}_0$, then, for stability, the resulting perturbed solution $x(t; \tilde{x}_0, t_0)$ is required to stay close to $x(t; x_0, t_0)$ for all $t \geq t_0$. In addition, for asymptotic stability, the error $x(t; \tilde{x}_0, t_0) \to x(t; x_0, t_0)$ is required to vanish as $t \to \infty$.

Definition A.1 The solution $x(t; x_0, t_0)$ of (A.1) is

- *Bounded*, if there exists a constant $B(x_0, t_0) > 0$, such that

$$|x(t; x_0, t_0)| < B(x_0, t_0), \quad \forall\, t \geq t_0; \tag{A.2}$$

- *Stable*, if for each $\epsilon > 0$ there exists a $\delta(\epsilon, t_0) > 0$, such that

$$|\tilde{x}_0 - x_0| < \delta, \quad |x(t; \tilde{x}_0, t_0) - x(t; x_0, t_0)| < \epsilon, \quad \forall\, t \geq t_0; \tag{A.3}$$

- *Attractive*, if there exists a $r(t_0) > 0$, and for each $\epsilon > 0$ and a $T(\epsilon, t_0) > 0$, such that

$$|\tilde{x}_0 - x_0| < r, \quad |x(t; \tilde{x}_0, t_0) - x(t; x_0, t_0)| < \epsilon, \quad \forall\, t \geq t_0 + T; \tag{A.4}$$

- *Asymptotically stable*, if it is stable and attractive; and

- *Unstable*, if it is not stable.

211

Uniform Stability [53]

Theorem A.1 Let $x = 0$ be an equilibrium point of (A.1) and $D = \{x \in \mathscr{R}^n \mid |x| < r$. Let $V : D \times \mathscr{R}^n \to \mathscr{R}_+$ be a continuously differentiable function such that $\forall t \geq 0$, $\forall x \in D$, such that

$$\gamma_1(|x|) \leq V(x, t) \leq \gamma_2(|x|) \tag{A.5}$$

$$\frac{\partial V}{\partial t} + \frac{\partial V}{\partial x} f(x, t) \leq -\gamma_3(|x|) \tag{A.6}$$

Then the equilibrium $x = 0$ is

- Uniformly stable, if γ_1 and γ_2 are class κ functions on $[0, r)$ and $\gamma_3(.) \geq 0$ on $[0, r)$;

- Uniformly asymptotically stable, if γ_1, γ_2 and γ_3 are class κ functions on $[0, r)$;

- Exponentially stable, if $\gamma_i(\rho) = k_i \rho^\alpha$ on $[0, r)$, $k_i > 0, \alpha > 0, i = 1, 2, 3$;

- Globally uniformly stable, if $D = R^n$, γ_1 and γ_2 are class κ_∞ functions, and $\gamma_3(.) \geq 0$ on $\mathscr{R}_+$;

- Globally uniformly asymptotically stable, if $D = \mathscr{R}^n$, γ_1 and γ_2 are class κ_∞ functions, and γ_3 is a class of κ function on $\mathscr{R}_+$; and

- Globally exponentially stable, if $D = \mathscr{R}^n$ and $\gamma_i(\rho) = k_i \rho^\alpha$ on $\mathscr{R}_+, k_i > 0, \alpha > 0, i = 1, 2, 3$.

Appendix B

LaSalle-Yoshizawa Theorem [53]

Theorem B.1 Let $x = 0$ be an equilibrium point of (A.1) and suppose f is locally Lipschitz in x uniformly in t. Let $V : \mathscr{R}^n \times \mathscr{R}_+ \to \mathscr{R}_+$ be a continuously differentiable function such that

$$\gamma_1(|x|) \leq V(x,t) \leq \gamma_2(|x|) \tag{B.1}$$

$$\dot{V} = \frac{\partial V}{\partial t} + \frac{\partial V}{\partial x} f(x,t) \leq -W(x) \leq 0 \tag{B.2}$$

$\forall\, t \geq 0,\ \forall\, x \in \mathscr{R}^n$, where γ_1 and γ_2 are class k_∞ functions and W is a continuous function. Then, all solutions of (A.1) are globally uniformly bounded and satisfy

$$lim_{t \to \infty} W(x(t)) = 0 \tag{B.3}$$

In addition, if $W(x)$ is positive definite, then the equilibrium $x = 0$ is globally uniformly asymptotically stable.

Barbalat Lemma [53]

Theorem B.2 Consider the function $f(t): \mathscr{R}_+ \to \mathscr{R}$. If $f(t)$ is uniformly continuous and $\lim_{t \to \infty} \int_0^\infty f(\tau)d\tau$ exists and is finite, then

$$\lim_{t \to \infty} f(t) = 0. \tag{B.4}$$

Corollary B.1 Consider the function $f(t): \mathscr{R}_+ \to R$. If $f(t)$, $\dot{f}(t) \in \mathscr{L}_\infty$ and $f(t) \in \mathscr{L}_p$ for some $p \in [1, \infty)$, then

$$\lim_{t \to \infty} f(t) = 0. \tag{B.5}$$

213

Appendix C

Parameter Projection [53]

Defining the following convex set

$$II_\epsilon = \{\hat{\theta} \in IR^p | P(\hat{\theta}) \leq \epsilon\}, \quad II = \{\hat{\theta} \in IR^p | P(\hat{\theta}) \leq 0\} \tag{C.1}$$

which is a union of the set II and an $O(\epsilon)$-boundary layer around it. Let us denote the interior of II_ϵ by II° and observe that $\nabla_{\hat{\theta}} P$ represents an outward normal vector at $\hat{\theta} \in \partial II_\epsilon$. The standard projection operator is

$$Proj\{\tau\} = \begin{cases} \tau & \hat{\theta} \in II^o \text{ or } \nabla_{\hat{\theta}} P^t \tau \leq 0 \\ \left(I - c(\hat{\theta})\Gamma \dfrac{\nabla_{\hat{\theta}} P \nabla_{\hat{\theta}} P^T}{\nabla_{\hat{\theta}} P^T \Gamma \nabla_{\hat{\theta}} P}\right)\tau & \hat{\theta} \in II_\epsilon / II^o \text{ and } \nabla_{\hat{\theta}} P^T \tau > 0 \end{cases} \tag{C.2}$$

$$c(\hat{\theta}) = min\{1, \frac{P(\hat{\theta})}{\epsilon}\} \tag{C.3}$$

where Γ belongs to the set G of all positive definite symmetric $p \times p$ matrices. It is helpful to note that $c(\partial II_\epsilon) = 1$.

Theorem C.1 The projection operator (C.2) has the following properties:

- The mapping $Proj: IR^p \times II_\epsilon \times G \to IR^p$ is locally Lipschiz in its arguments $\tau, \hat{\theta}, \Gamma$.

- $Proj\{\tau\}^T \Gamma^{-1} Proj\{\tau\} \leq \tau^T \Gamma^{-1} \tau, \quad \forall \hat{\theta} \in II_\epsilon$.

- Let $\Gamma(t), \tau(t)$ be continuously differentiable and $\dot{\hat{\theta}} = Proj\{\tau\}$, $\hat{\theta}(0) \in II_\epsilon$. Then, on its domain of definition, the solution $\hat{\theta}(t)$ remains in II_ϵ.

- $-\tilde{\theta}^T \Gamma^{-1} Proj\{\tau\} \leq -\tilde{\theta}^T \Gamma^{-1} \tau, \forall \hat{\theta} \in II_\epsilon, \theta \in II$.

Appendix D

Non-smooth Stability Analysis

Consider the following class of differential equations

$$\dot{x}(t) = X(x(t)) \tag{D.1}$$

where $X : \mathbb{R}^d \to \mathbb{R}^d$ is measurable but discontinuous, the existence of a continuously differentiable solution in the classical sense is not guaranteed. In this case, the Filippov solution [16] can be considered.

Definition D.1 (Filippov solution) [26]
Let $\mathcal{B}(\mathbb{R}^d)$ denote the collection of all subsets of $\mathbb{R}^d$. The Filippov set-valued map $\mathcal{F}[X] : \mathbb{R}^d \to \mathcal{B}(\mathbb{R}^d)$ is defined by

$$\mathcal{F}[X](x) \triangleq \cap_{\varsigma > 0} \cap_{\mu(\varsigma)=0} \overline{co}\{X(\mathbf{B}(\mathbb{R}^d, \varsigma) \setminus o)\}, x \in \mathbb{R}^d \tag{D.2}$$

where o is the set of x which makes $X(x)$ discontinuous, $\mathbf{B}(\mathbb{R}^d, \varsigma)$ is a open ball of radius ς centered at x, $\overline{co}$ represents the convex closure, while $\cap_{\mu(\varsigma)=0}$ denotes the intersection over all sets o of Lebesgue measure zero. Then, the Filippov solutions are defined as the absolutely continuous curves satisfying the differential inclusion

$$\dot{x}(t) \in \mathcal{F}[X](x) \tag{D.3}$$

The existence of the Filippov solution is ensured by the following Lemma.

Lemma D.1 [16] Assume $X : \mathbb{R}^d \to \mathbb{R}^d$ is measurable and locally essentially bounded, i.e. bounded in any bounded neighborhood of every point of definition excluding the sets of measure zero. Then for all $x_0 \in \mathbb{R}^d$, there exists a Filippov solution to (D.1) with the initial condition $x(0) = x_0$.

Lemma D.2 [15] Let $x(\cdot)$ be a Filippov solution to $\dot{x}(t) = X(x(t))$ on an interval containing t, and $V : \mathbb{R}^d \to \mathbb{R}$ be a Lipschitz and regular function. Then $V(x(t))$

is absolutely continuous and $\frac{d}{dt}V(x(t))$ exists almost everywhere such that

$$\frac{d}{dt}V(x(t)) \in \dot{\tilde{V}}(x), \quad for\ a.e.\ t \geq 0, \tag{D.4}$$

where

$$\dot{\tilde{V}}(x) = \cap_{\nu \in \partial V(x(t))} \nu^T \mathcal{F}[X](x) \tag{D.5}$$

References

[1] S. Anand, B. G. Fernandes, and J. Guerrero. Distributed control to ensure proportional load sharing and improve voltage regulation in low-voltage DC microgrids. *IEEE Transactions on Power Electronics*, 28(4):1900–1913, 2013.

[2] T. S. Andersen and R. Kristiansen. Adaptive backstepping control for a fully-actuated rigid-body in a dual-quaternion framework. In *2019 IEEE 58th Conference on Decision and Control (CDC)*. IEEE, 2019.

[3] A. F. Anton Selivanov and and D. Liberzon. Adaptive control of passifiable linear systems with quantized measurements and bounded disturbances. *Systems and Control Letters*, 88:62–67, 2016.

[4] S. Augustine, M. K. Mishra, and N. Lakshminarasamma. Adaptive droop control strategy for load sharing and circulating current minimization in low-voltage standalone DC microgrid. *IEEE Transactions on Sustainable Energy*, 6(1):132–141, 2015.

[5] A. Benallegue, Y. Chitour, and A. Tayebi. Adaptive attitude tracking control of rigid body systems with unknown inertia and gyro-bias. *IEEE Transactions on Automatic Control*, 63(11):3986–3993, 2018.

[6] D. S. Bernstein. Matrix mathematics: Theory, facts, and formulas, Princeton University Press, 2009.

[7] V. A. Bondarko and A. L. Fradkov. Adaptive stabilization of linear systems through a two-way channel with limited capacity. *IFAC-PapersOnLine*, 91(13):164–168, 2016.

[8] V. A. Bondarko and A. L. Fradkov. Adaptive stabilisation of discrete lti plant with bounded disturbances via finite capacity channel. *International Journal of Control*, 91(11):2451–2459, 2018.

[9] A.-C. Braitor, G. Konstantopoulos, and V. Kadirkamanathan. Current-limiting droop control design and stability analysis for paralleled boost converters in DC microgrids. *IEEE Transactions on Control System Technology*, 29(1):385–394, 2020.

[10] H. Cai and G. Hu. Distributed nonlinear hierarchical control of ac microgrid via unreliable communication. *IEEE Transactions on Smart Grid*, 9(4):2429–2441, 2018.

[11] R. Carli, F. Bullo, and S. Zampieri. Quantized average consensus via dynamic coding/decoding schemes. *International Journal of Robust and Nonlinear Control: IFAC-Affiliated Journal*, 20(2):156–175, 2010.

[12] F. Ceragiolia, C. D. Persis, and P. Frascaa. Discontinuities and hysteresis in quantized average consensus. *Automatica*, 47(9):1916–1928, 2011.

[13] N. A. Chaturvedi, A. K. Sanyal, and N. H. McClamroch. Rigid-body attitude control. *IEEE Control Systems Magazine*, 31(3):30–51, 2011.

[14] F. Chen, R. Jiang, K. Zhang, B. Jiang, and G. Tao. Robust backstepping sliding-mode control and observer-based fault estimation for a quadrotor uav. *IEEE Transactions on Industrial Electronics*, 63(6):5044–5056, 2016.

[15] F. H. Clarke. *Optimization and Nonsmooth Analysis*. SIAM, 1990.

[16] J. Cortes. Discontinuous dynamical systems. *IEEE Control Systems magzine*, 28(3):36–73, 2008.

[17] D. F. Coutinho, M. Fu, and C. E. de Souza. Input and output quantized feedback linear systems. *IEEE Transactions on Automatic Control*, 55(3):761–766, 2010.

[18] A. Datta and P. Ioannou. Decentralized indirect adaptive control of interconnected systems. *International Journal of Adaptive Control and Signal Processing*, 5:259–281, 1991.

[19] A. Datta and P. Ioannou. Decentralized adaptive control. *Advances in Control and Dynamic systems*, page C. T. Leondes(Ed.) Academic, 1992.

[20] C. De Persis and F. Mazenc. Stability of quantized time-delay nonlinear systems: A Lyapunov-Krasowskii functional approach. *Mathematics of Control, Signals, and Systems*, 21:4337–370, 2010.

[21] Z. Ding. Adaptive asymptotic tracking of nonlinear output feedback systems under unknown bounded disturbances. *System and Science*, 24:47–59, 1998.

[22] M. Dong, L. Li, Y. Nie, D. Song, and J. Yang. Stability analysis of a novel distributed secondary control considering communication delay in DC microgrids. *IEEE Transactions on Smart Grid*, 10(6):6690–6700, 2019.

[23] N. Elia and S. K. Mitter. Stabilization of linear systems with limited information. *IEEE transactions on Automatic Control*, 46(9):1384–1400, 2001.

[24] F. Fagnani and S. Zampieri. Quantized stabilization of linear systems: complexity versus performance. *IEEE Transactions on automatic control*, 49(9):1534–1548, 2004.

[25] B. Fan, J. Peng, J. Duan, Q. Yang, and W. Liu. Distributed control of multiple-bus microgrid with paralleled distributed generators. *IEEE/CAA Journal of Automatica Sinica*, 6(3):676–684, 2019.

[26] A. F. Filippov. *Differential Equations with Discontinuous Righthand Sides*. Springer, 1988.

[27] T. I. Fossen. *Marine Control Systems: Guidance, Navigation, and Control of Ships, Rigs and Underwater Vehicles*. Marine Cybernetics AS, Trondheim, Norway, 2002.

[28] E. Fridman and M. Dambrine. Control under quantization, saturation and delay: An lmi approach. *Automatica*, 45(10):2258–2264, 2009.

[29] M. Fu and L. Xie. The sector bound approach to quantized feedback control. *IEEE Transactions on Automatic control*, 50(11):1698–1711, 2005.

[30] J. M. Guerrero, J. C. Vasquez, J. Matas, L. G. D. Vicuña, and M. Castilla. Hierarchical control of droop-controlled ac and DC microgrids-a general approach toward standardization. *IEEE Transactions on Industrial Electronics*, 58(1):158–172, 2011.

[31] F. Guo, Q. Xu, C. Wen, L. Wang, and P. Wang. Distributed secondary control for power allocation and voltage restoration in islanded dc microgrids. *IEEE Transactions on Sustainable Energy*, 9(4):1857–1869, 2018.

[32] R. Han, M. Tucci, A. Martinelli, J. M. Guerrero, and G. Ferrari-Trecate. Stability analysis of primary plug-and-play and secondary leader-based controllers for DC microgrid clusters. *IEEE Transactions on Power Systems*, 34(3):1780–1800, 2018.

[33] T. Hayakawaa, H. Ishii, and K. Tsumurac. Adaptive quantized control for linear uncertain discrete-time systems. *Automatica*, 45:692–700, 2009.

[34] T. Hayakawaa, H. Ishii, and K. Tsumurac. Adaptive quantized control for nonlinear uncertain systems. *Systems and Control Letters*, 58:625–632, 2009.

[35] D. J. Hill, C. Wen, and G. C. Goodwin. Stability analysis of decentralized robust adaptive control. *System and Control Letters*, 11:277–284, 1988.

[36] B. Huang, B. Zhou, S. Zhang, and C. Zhu. Adaptive prescribed performance tracking control for underactuated autonomous underwater vehicles with input quantization. *Ocean Engineering*, 221, 2021.

[37] J. Huang, W. Wang, C. Wen, J. Zhou, and G. Li. Distributed adaptive leader-follower and leaderless consensus control of a class of strict-feedback nonlinear systems: a unified approach. *Automatica*, 118(109021):1–9, 2020.

[38] P. Ioannou. Decentralized adaptive control of interconnected systems. *IEEE Transactions on Automatic Control*, 31:291–298, 1986.

[39] P. Ioannou and P. Kokotovic. An asymptotic error analysis of identifiers and adaptive observers in the presence of parasitics. *IEEE Transactions on Automatic Control*, 27:921–927, 1982.

[40] P. Ioannou and P. Kokotovic. Robust redesign of adaptive control. *IEEE Transactions on Automatic Control*, 29:202–211, 1984.

[41] P. Ioannou and P. Kokotovic. Decentralized adaptive control of interconnected systems with reduced-order models. *Automatica*, 21:401–412, 1985.

[42] P. Ioannou and K. Tsakalis. A robust direct adaptive controller. *IEEE Transactions on Automatic Control*, 31:1033–1043, 1986.

[43] P. A. Ioannou and P. V. Kokotovic. *Adaptive systems with reduced models*. Springer-Verlag, Berlin, Germany, 1983.

[44] H. Ishii and T. Başar. Remote control of LTI systems over networks with state quantization. *Systems & Control Letters*, 54(1):15–31, 2005.

[45] H. Ishii and B. Francis. *Limited Data Rate in Control Systems with Network*. Springer, Berlin, Germany, 2002.

[46] A. Isidori. *Nonlinear Control Systems: An Introduction*. Springer-Verlag, Berlin, Germany, 1989.

[47] R. Iskakov, A. Albu-Schaeffer, M. Schedl, G. Hirzinger, and V. Lopota. Influence of sensor quantization on the control performance of robotics actuators. In *2007 IEEE/RSJ International Conference on Intelligent Robots and Systems*, pages 1085–1092. IEEE, 2007.

[48] Z. P. Jiang. Decentralized and adaptive nonlinear tracking of large-scale systems via output feedback. *IEEE Transactions on Automatic Control*, 45:2122–2128, 2000.

[49] K. H. Johansson, A. Speranzon, and S. Zampieri. On quantization and communication topologies in multi-vehicle rendezvous. *IFAC Proceedings Volumes*, 38(1):109–114, 2005.

[50] R. E. Kalman. Nonlinear aspects of sampled-data control systems. In *Proc. Symp. Nonlinear Circuit Analysis VI, 1956*, pages 273–313, 1956.

[51] H. Khalil. *Nonlinear systems*. Person Education International, New Jersey, USA, 1992.

[52] H. K. Khalil. *Nonlinear Control*. Pearson, 2015.

[53] M. Krstic, I. Kanellakopoulos, and P. V. Kokotovic. *Nonlinear and Adaptive Control Design*. Wiley, New York, 1995.

[54] G. Lai, Z. Liu, Y. Zhang, C. L. P. Chen, and S. Xie. Asymmetric actuator backlash compensation in quantized adaptive control of uncertain networked nonlinear systems. *IEEE transactions on neural networks and learning systems*, 28(2):294–307, 2015.

[55] R. Lasseter, A. Akhil, C. Marnay, J. Stephens, J. Dagle, R. Guttromson, A. Meliopoulous, R. Yinger, and J. Eto. White paper on integration of distributed energy resources: The certs microgrid concept. *Consortium for Electric Reliability Technology Solutions*, pages 1–27, 2003.

[56] T. Lee. Robust adaptive attitude tracking on SO(3) with an application to a quadrotor UAV. *IEEE Transactions on Control Systems Technology*, 21(5):1924–1930, 2013.

[57] H. Lei and W. Lin. Universal adaptive control of nonlinear systems with unknown growth rate by output feedback. *Automatica*, 42(10):1783–1789, 2006.

[58] G. Li and Y. Lin. Adaptive output feedback control for a class of nonlinear systems with quantised input and output. *International Journal of Control*, 90(2):239–248, 2017.

[59] Y. Li and G. Yang. Adaptive asymptotic tracking control of uncertain nonlinear systems with input quantization and actuator faults. *Automatica*, 72:177–185, 2016.

[60] D. Liberzon. Hybrid feedback stabilization of systems with quantized signals. *Automatica*, 39:1543–1554, 2003.

[61] D. Liberzon and J. Hespanha. Stabilization of nonlinear systems with limited information feedback. *IEEE Transactions on Automatic Control*, 50:910–915, 2005.

[62] J. Liu and N. Elia. Quantized feedback stabilization of non-linear affine systems. *International Journal of Control*, 77:239–249, 2004.

[63] K. Liu, E. Fridman, and K. H. Johansson. Dynamic quantization of uncertain linear networked control systems. *Automatica*, 59:248–255, 2015.

[64] L.-J. Liu, J. Zhou, C. Wen, and X. Zhao. Robust adaptive tracking control of uncertain systems with time-varying input delays. *International Journal of Systems Science*, 48(16):3440–3449, 2017.

[65] T. Liu and Z. Jiang. Event-triggered control of nonlinear systems with state quantization. *IEEE Transactions on Automatic Control*, 64(2):797–803, 2019.

[66] T. Liu, Z. P. Jiang, and D. J. Hill. Quantized stabilization of strict-feedback nonlinear systems based on ISS cyclic-small-gain theorem. *Mathematics of Control, Signals, and Systems*, 24, issue 1-2:75–110, 2012.

[67] T. Liu, Z.-P. Jiang, and D. J. Hill. A sector bound approach to feedback control of nonlinear systems with state quantization. *Automatica*, 48:145–152, 2012.

[68] X.-K. Liu, H. He, Y.-W. Wang, Q. Xu, and F. Guo. Distributed hybrid secondary control for a DC microgrid via discrete-time interaction. *IEEE Transactions on Energy Conversion*, 33(4):1865–1875, 2018.

[69] X.-K. Liu, Y.-W. Wang, P. Lin, and P. Wang. Distributed supervisory secondary control for a DC microgrid. *IEEE Transactions on Energy Convers*, 35(4):1736–1746, 2020.

[70] R. Lozano and B. Brogliato. Adaptive control of a simple nonlinear system without a priori information on the plant parameters. *IEEE Transactions on Automatic Control*, 1992.

[71] R. Marino and P. Tomei. *Nonlinear Control Design: Geometric, Adaptive and Robust*. Prentice Hall, New York, 1995.

[72] R. H. Middleton and G. C. Goodwin. Adaptive control of time-varying linear systems. *IEEE Transactions on Automatic Control*, 33:150–155, 1988.

[73] R. H. Middleton and G. C. Goodwin. Indirect adaptive output-feedback control of a class on nonlinear systems. In *Proceedings of the 29th IEEE Conference on Decision and Control*, volume 5, pages 2714–2719, 1990.

[74] R. H. Middleton, G. C. Goodwin, D. J. Hill, and D. Q. Mayne. Design issues in adaptive control. *IEEE Transactions on Automatic Control*, 33:50–58, 1988.

[75] G. Nair and R. Evans. Stabilizability of stochastic linear systems with finite feedback data rates. *SIAM Journal on Control and Optimization*, 43:413–436, 2004.

[76] A. Nedic, A. Olshevsky, A. Ozdaglar, and J. N. Tsitsiklis. On distributed averaging algorithms and quantization effects. *IEEE Transactions on automatic control*, 54(11):2506–2517, 2009.

[77] C. Peng, J. Li, and M. Fei. Resilient event-triggering h_∞ load frequency control for multi-area power systems with energy-limited dos attacks. *IEEE Transactions on Power Systems*, 32(5):4110–4118, 2017.

[78] J. Peng, B. Fan, J. Duan, Q. Yang, and W. Liu. Adaptive decentralized output-constrained control of single-bus DC microgrids. *IEEE/CAA Journal of Automatica Sinica*, 6(2):424–432, 2019.

[79] J. Peng, B. Fan, Q. Yang, and W. Liu. Fully distributed discrete-time control of dc microgrids with zip loads. *IEEE Systems Journal*, 2020.

[80] C. D. Persis. Robust stabilization of nonlinear systems by quantized and ternary control. *Systems & Control Letters*, 58:602–609, 2009.

[81] C. D. Persis and A. Isidori. Stabilizability by state feedback implies stabilizability by encoded state feedback. *Systems and Control Letters*, 53:249–258, 2004.

[82] B. Picasso and A. Bicchi. On the stabilization of linear systems under assigned i/o quantization. *IEEE Transactions on Automatic Control*, 52(10):1994–2000, 2007.

[83] M. M. Polycarpou. Stable adaptive neural control scheme for nonlinear system. *IEEE Transactions on Automatic Control*, 41(3):447–451, 1996.

[84] J. B. Pomet and L. Praly. Adaptive nonlinear regulation: estimation from the lyapunov equation. *IEEE Transactions on Automatic Control*, 37:729–740, 1992.

[85] L. Praly. Towards a globally stable direct adaptive control scheme for not necessarily minimum phase systems. *IEEE Transactions on Automatic Control*, 29:946–949, 1984.

[86] L. Praly. Lyapunov design of stabilizing controllers for cascaded systems. *IEEE Transactions on Automatic Control*, 36:1177–1181, 1991.

[87] L. Praly. Towards an adaptive regulator: Lyapunov design with a growth condition. In *Proceedings of the 30th IEEE Conference on Decision and Control*, volume 2, pages 1094–1099, 1991.

[88] L. Praly and B. d'Andrea Novel; J. M. Coron. Lyapunov design of stabilizing controllers. In *Proceedings of the 28th IEEE Conference on Decision and Control*, volume 2, pages 1047–1052, 1989.

[89] C. Qian and W. Lin. Output feedback control of a class of nonlinear systems: a nonseparation principle paradigm. *IEEE Transactions on Automatic Control*, 47(10):1710–1715, 2002.

[90] R. Schlanbusch. *Control of rigid bodies*. PhD thesis, NTNU - Norwegian University of Science and Technology, 2012.

[91] R. Schlanbusch, A. Loria, R. Kristiansen, and P. J. Nicklasson. PD+ attitude control of rigid bodies with improved performance. In *49th IEEE Conference on Decision and Control*, pages 7069–7074, 2010.

[92] S. M. Schlanbusch and J. Zhou. Adaptive backstepping control of a 2-dof helicopter. In *Proceedings of the IEEE 7th International Conference on Control, Mechatronics and Automation*, 2019.

[93] S. M. Schlanbusch and J. Zhou. Adaptive backstepping control of a 2-DOF helicopter system with uniform quantized inputs. In *IECON 2020 The 46th Annual Conference of the IEEE Industrial Electronics Society*, pages 88–94, 2020.

[94] H. Sun, L. Hou, G. Zong, and X. Yu. Fixed-time attitude tracking control for spacecraft with input quantization. *IEEE Transactions on Aerospace and Electronic Systems*, 55(1):124–134, 2018.

[95] H. Sun, N. Hovakimyan, and T. Basar. L_1 adaptive controller for systems with input quantization. In *American Control Conference*, pages 253–258, Baltimore, USA, June 30-July 02 2010.

[96] R. Sun, A. Shan, C. Zhang, J. Wu, and Q. Jia. Quantized fault-tolerant control for attitude stabilization with fixed-time disturbance observer. *Journal of Guidance, Control, and Dynamics*, 44(2):1–7, 2020.

[97] G. Tao and P. Ioannou. Model reference adaptive control for plants with unknown relative degree. *IEEE Transactions on Automatic Control*, 38:976 – 982, 1993.

[98] S. Tatikonda and S. Mitter. Control under communication constraints. *IEEE Transactions on Automatic Control*, 49:1056–1068, 2004.

[99] L. Vu and D. Liberzon. Supervisory control of uncertain systems with quantized information. *International Journal of Adaptive Control and Signal Process*, 26:739–756, 2012.

[100] C. Wang, C. Wen, Q. Hu, W. Wang, and X. Zhang. Distributed adaptive containment control for a class of nonlinear multiagent systems with input quantization. *IEEE transactions on neural networks and learning systems*, 29(6):2419–2428, 2017.

[101] C. Wang, C. Wen, Y. Lin, and W. Wang. Decentralized adaptive tracking control for a class of interconnected nonlinear systems with input quantization. *Automatica*, 81:359–368, 2017.

[102] T. Wang and J. Huang. Leader-following adaptive consensus of multiple uncertain rigid body systems over jointly connected networks. *Unmanned Systems*, 08(02):85–93, 2020.

[103] W. Wang, C. Wen, J. Huang, and Z. Li. Hierarchical decomposition based consensus tracking for uncertain interconnected systems via distributed adaptive output feedback control. *IEEE Transactions on Automatic Control*, 61(7):1938–1945, 2015.

[104] W. Wang, C. Wen, J. Huang, and J. Zhou. Adaptive consensus of uncertain nonlinear systems with event triggered communication and intermittent actuator faults. *Automatica*, 111:108667, 2020.

[105] Y. Wang, L. He, and C. Huang. Adaptive time-varying formation tracking control of unmanned aerial vehicles with quantized input. *ISA Transactions*, 85:76–83, 2019.

[106] C. Wen. A robust adaptive controller with minimal modifications for discrete time varying systems. In *Proceedings of the 31st IEEE Conference on Decision and Control*, volume 2, pages 2132 – 2136, 1992.

[107] C. Wen. Robustness of a simple indirect continuous time adaptive controller in the presence of bounded disturbances. In *Proceedings of the 31st IEEE Conference on Decision and Control*, volume 3, pages 2762 – 2766, 1992.

[108] C. Wen. Decentralized adaptive regulation. *IEEE Transactions on Automatic Control*, 39:2163–2166, 1994.

[109] C. Wen. Indirect robust totally decentralized adaptive control of continuous-time interconnected systems. *IEEE Transactions on Automatic Control*, 40:1122–1126, 1995.

[110] C. Wen and D. J. Hill. Robustness of adaptive control without deadzones, data normalization or persistence of excitation. *Automatica*, 25:943–947, 1989.

[111] C. Wen and D. J. Hill. Adaptive linear control of nonlinear systems. *IEEE Transactions on Automatic Control*, 35:1253 – 1257, 1990.

[112] C. Wen and D. J. Hill. Decentralized adaptive control of lineartime varying systems. In *Proceedings of 11th IFAC World Cengress Automatica control*, Tallinn, U.S.S.R, 1990.

[113] C. Wen and D. J. Hill. Global boundedness of discrete-time adaptive control just using estimator projection. *Automatica*, 28:1143–1157, 1992.

[114] C. Wen and D. J. Hill. Globally stable discrete time indirect decentralized adaptive control systems. In *Proceedings of the 31st IEEE Conference on Decision and Control*, volume 1, pages 522 – 526, 1992.

[115] C. Wen and Y. C. Soh. Decentralized adaptive control using integrator backstepping. *Automatica*, 33:1719–1724, 1997.

[116] C. Wen, Y. Zhang, and Y. C. Soh. Robustness of an adaptive backstepping controller without modification. *Systems & Control Letters*, 36:87–100, 1999.

[117] C. Wen and J. Zhou. Decentralized adaptive stabilization in the presence of unknown backlash-like hysteresis. *Automatica*, 43:426–440, 2007.

[118] C. Wen, J. Zhou, Z. Liu, and H. Su. Robust adaptive control of uncertain nonlinear systems in the presence of input saturation and external disturbance. *IEEE Transactions on Automatic Control*, 56:1672–1678, 2011.

[119] C. Wen, J. Zhou, and W. Wang. Decentralized adaptive backstepping stabilization of interconnected systems with dynamic input and output interactions. *Automatica*, 45:55–67, 2009.

[120] J. T.-Y. Wen and K. Kreutz-Delgado. The attitude control problem. *IEEE Transactions on Automatic Control*, 36(10):1148–1162, 1991.

[121] B. Wu and X. Cao. Robust attitude tracking control for spacecraft with quantized torques. *IEEE Transactions on Aerospace and Electronic Systems*, 54(2):1020–1028, 2017.

[122] L. Xing, Y. Mishra, F. Guo, P. Lin, Y. Yang, G. Ledwich, and Y.-C. Tian. Distributed secondary control for current sharing and voltage restoration in DC microgrid. *IEEE Transactions on Smart Grid*, 11(3):2487–2497, 2020.

[123] L. Xing, Y. Mishra, Y.-C. Tian, G. Ledwich, H. Su, C. Peng, and M. Fei. Dual-consensus-based distributed frequency control for multiple energy storage systems. *IEEE Transactions on Smart Grid*, 10(6):6396– 6403, 2019.

[124] L. Xing, C. Wen, H. Su, J. Cai, and L. Wang. A new adaptive control scheme for uncertain nonlinear systems with quantized input signal. *Journal of the Franklin Institute*, 352:5599–5610, 2015.

[125] L. Xing, C. Wen, H. Su, G. Lai, and Z. Li. Robust adaptive output feedback control for uncertain nonlinear systems with quantized input. *International Journal of Robust and Nonlinear Control*, 27:1999–2016, 2017.

[126] L. Xing, C. Wen, L. Wang, Z. Liu, and H. Su. Adaptive output feedback regulation for a class of nonlinear systems subject to input and output quantization. *Journal of the Franklin Institute*, 354:6536–6549, 2017.

[127] L. Xing, C. Wen, Y. Zhu, H. Su, and Z. Liu. Output feedback control for uncertain nonlinear systems with input quantization. *Automatica*, 65:191–202, 2016.

[128] Y. Yan and S. Yu. Sliding mode tracking control of autonomous underwater vehicles with the effect of quantization. *Ocean Engineering*, 151:322–328, 2018.

[129] Z. Yang, Y. Hong, Z. P. Jiang, and X. Wang. Quantized feedback stabilization of hybrid impulsive control systems. In *Joint 48th IEEE Conference on Decision and Control and 28th Chinese Control Conference*, pages 3903–3908, Shanghai, P.R. China, 2009.

[130] L. Zhang, Z. Ning, and W. X. Zheng. Observer-based control for piecewise-affine systems with both input and output quantization. *IEEE Transactions on Automatic Control*, 62(11):5858–5865, 2016.

[131] L. Zhang, Z. Ning, and W. X. Zheng. Observer-based control for piecewise-affine systems with both input and output quantization. *IEEE Transactions on Automatic Control*, 62(11):5858–5865, 2017.

[132] X. Zhang, Y. Wang, C. Wang, C.-Y. Su, Z. Li, and X. Chen. Adaptive estimated inverse output-feedback quantized control for piezoelectric positioning stage. *IEEE transactions on cybernetics*, 49(6):2106–2118, 2018.

[133] X. Zhang, Y. Wang, G. Zhu, X. Chen, Z. Li, C. Wang, and C.-Y. Su. Compound adaptive fuzzy quantized control for quadrotor and its experimental verification. *IEEE Transactions on Cybernetics*, 51(3):1121–1133, 2020.

[134] Y. Zhang, C. Wen, and Y. C. Soh. Robust adaptive control of uncertain discrete-time systems. *Automatica*, 35:321–329, 1999.

[135] Y. Zhang, C. Wen, and Y. C. Soh. Adaptive backstepping control design for systems with unknown high-frequency gain. *IEEE Transactions on Automatic Control*, 45:2350–2354, 2000.

[136] Y. Zhang, C. Wen, and Y. C. Soh. Discrete-time robust adaptive control for nonlinear time-varying systems. *IEEE Transactions on Automatic Control*, 45:1749–1755, 2000.

[137] Y. Zhang, C. Wen, and Y. C. Soh. Robust decentralized adaptive stabilization of interconnected systems with guaranteed transient performance. *Automatica*, 36:907–915, 2000.

[138] Y. Zhang, C. Wen, and Y. C. Soh. Robust adaptive control of nonlinear discrete-time systems by backstepping without overparameterization. *Automatica*, 37:551–558, 2001.

[139] J. Zhou. Decentralized adaptive backstepping stabilization of interconnected systems with input time delays in dynamic interactions. *International Journal of Adaptive Control and Signal Processing*, 26(4):285–301, 2012.

[140] J. Zhou and M. Krstic. Adaptive predictor control for stabilizing pressure in a managed pressure drilling system under time-delay. *Journal of Process Control*, 40:106–118, 2016.

[141] J. Zhou and C. Wen. *Adaptive Backstepping Control of Uncertain Systems: Nonsmooth Nonlinearities, Interactions or Time-Variations*. Springer, 2008.

[142] J. Zhou and C. Wen. Adaptive backstepping control of uncertain nonlinear systems with input quantization. In *52nd IEEE Conference on Decision and Control*, pages 5571–5576, Florence, Italy, 2013.

[143] J. Zhou, C. Wen, and W. Cai. Adaptive control of a base isolated system for protection of building structures. *Journal of Vibration and Acoustics, ASME*, 128:261–268, 2006.

[144] J. Zhou, C. Wen, and T. Li. Adaptive output feedback control of uncertain non-linear systems with hysteresis nonlinearity. *IEEE Transactions on Automatic Control*, 57(10):2627–2633, 2012.

[145] J. Zhou, C. Wen, and W. Wang. Adaptive control of uncertain nonlinear systems with quantized input signal. *Automatica*, 95:152–162, 2018.

[146] J. Zhou, C. Wen, W. Wang, and F. Yang. Adaptive backstepping control of nonlinear uncertain systems with quantized states. *IEEE Transactions on Automatic Control*, 64(11):4756–4763, 2019.

[147] J. Zhou, C. Wen, and G. Yang. Adaptive backstepping stabilization of nonlinear uncertain systems with quantized input signal. *IEEE Transactions on Automatic Control*, 59:460–464, 2014.

[148] J. Zhou, C. Wen, and Y. Zhang. Adaptive backstepping control of a class of uncertain nonlinear systems with unknown backlash-like hysteresis. *IEEE Transactions on Automatic Control*, 49:1751–1757, 2004.

[149] J. Zhou, C. Wen, and Y. Zhang. Adaptive output control of a class of time-varying uncertain nonlinear systems. *Journal of Nonlinear Dynamics and System Theory*, 3:285–298, 2005.

[150] J. Zhou, C. Wen, and Y. Zhang. Adaptive output control of nonlinear systems with uncertain dead-zone nonlinearity. *IEEE Transactions on Automatic Control*, 51:504–511, 2006.

[151] J. Zhou, Y. Xu, H. Sun, L. Wang, and M.-Y. Chow. Distributed event-triggered h_∞ consensus based current sharing control of DC microgrids considering uncertainties. *IEEE Transactions on Industrial Informatics*, 16(12):7413–7425, 2020.

[152] J. Zhou, C. Zhang, and C. Wen. Robust adaptive output control of uncertain nonlinear plants with unknown backlash nonlinearity. *IEEE Transactions on Automatic Control*, 52(3):503–509, 2007.

[153] W. Zhuang, C. Wen, J. Zhou, Z. Liu, and H. Su. Event-triggered robust adaptive control for discrete time uncertain systems with unmodelled dynamics and disturbances. *IET Control Theory & Applications*, 13(18):3124–3131, 2019.

Index